VOLUME NINE

ANNUAL REPORTS IN
COMPUTATIONAL
CHEMISTRY

VOLUME NINE

Annual Reports in COMPUTATIONAL CHEMISTRY

Edited by

Ralph A. Wheeler
Department of Chemistry and Biochemistry,
Duquesne University,
600 Forbes Avenue,
Pittsburgh,
PA 15282-1530

Sponsored by the Division of Computers in Chemistry of the American Chemical Society

ELSEVIER   AMSTERDAM • BOSTON • HEIDELBERG • LONDON • NEW YORK • OXFORD
PARIS • SAN DIEGO • SAN FRANCISCO • SINGAPORE • SYDNEY • TOKYO

Elsevier
Radarweg 29, PO Box 211, 1000 AE Amsterdam, The Netherlands
The Boulevard Langford Lane, Kidlingfon, Oxford, OX5 1GB, UK

First edition 2013

Library of Congress Cataloging-in-Publication Data
A catalogue record for this book is available from the Library of Congress

British Library Cataloguing in Publication Data
A catalog record for this book is available from the British Library

ISBN: 978-0-444-62672-1
ISSN: 1574-1400

For information on all Elsevier publications
visit our website at store.elsevier.com

Printed and bound by CPI Group (UK) Ltd, Croydon, CR0 4YY

Transferred to digital print 2013

Working together
to grow libraries in
developing countries

www.elsevier.com • www.bookaid.org

CONTENTS

CONTRIBUTORS

Wesley D. Allen
Department of Chemistry and Center for Computational Chemistry, University of Georgia, Athens, Georgia, USA

Jenna A. Bilbrey
Department of Chemistry and Center for Computational Chemistry, University of Georgia, Athens, Georgia, USA

Chia-en A. Chang
Department of Chemistry, University of California, Riverside, California, USA

John M. Herbert
Department of Chemistry and Biochemistry, The Ohio State University, Columbus, Ohio, USA

Yu-ming M. Huang
Department of Chemistry, University of California, Riverside, California, USA

Leif D. Jacobson
Department of Chemistry, Yale University, New Haven, Connecticut, USA

Ka Un Lao
Department of Chemistry and Biochemistry, The Ohio State University, Columbus, Ohio, USA

Ryan M. Richard
Department of Chemistry and Biochemistry, The Ohio State University, Columbus, Ohio, USA

PREFACE

Annual Reports in Computational Chemistry (ARCC) is published and distributed by Elsevier, sponsored by the American Chemical Society's Division of Computers in Chemistry (COMP), and focuses on providing concise, timely reviews of topics important to researchers in computational chemistry. All members in good standing of the COMP Division receive a copy of the *ARCC* as part of their member benefits. The Executive Committee of the COMP Division is very pleased that previous volumes have received an enthusiastic response from readers, and we expect to continue the tradition by delivering high-quality volumes of *ARCC*. To ensure that you receive future instalments of this series, please join the Division as described on the COMP Web site at http://www.acscomp.org.

Volume 9 is quite slim but contains high-quality contributions. Section A (Quantum Chemistry, edited by Gregory S. Tschumper) leads off with a review of new steric descriptors that extend the traditional solid angles and Tolman cone angles to account for ligand flexibility. The next contribution describes recently developed fragment-based quantum chemistry methods designed to achieve chemical accuracy. Section B (Bioinformatics, edited by Wei Wang) contains a contribution describing phosphopeptide binding to modular protein domains and their role in signalling events important in cellular functions. To provide easy identification of past reports, we plan to continue the practice of cumulative indexing of both this and past editions.

This and past volumes of *ARCC* have been assembled entirely by volunteers to produce a high-quality scientific publication at the lowest cost possible. The editor and the COMP Executive Committee extend our gratitude to the many people who have given their time so generously to make this edition of *ARCC* possible. The authors of each of this year's contributions and the section editors have graciously dedicated significant amounts of their time to make this volume successful. This year's edition could not have been assembled without the help of Shellie Bryant and Kate Newell of Elsevier. Thank you one and all for your hard work, time, and contributions.

We trust that you will continue to find *ARCC* to be interesting and valuable. We are actively planning the tenth volume to rival past volumes in size and quality, with several new sections and section editors. We welcome input from our readers about future topics, so please contact the editor to make suggestions and/or to volunteer as a contributor.

Sincerely,

Ralph A. Wheeler, Editor

Quantum Chemistry

Section Editor: Gregory S. Tschumper

Department of Chemistry and Biochemistry,
University of Mississippi, University,
MS, USA

CHAPTER ONE

Ligand Steric Descriptors

Jenna A. Bilbrey, Wesley D. Allen[1]

Department of Chemistry and Center for Computational Chemistry, University of Georgia, Athens, Georgia, USA
[1]Corresponding author: e-mail address: wdallen@uga.edu

Contents

Abstract

Descriptors for the steric size of transition metal-bound ligands suffer from coarse approximations and complicated correction schemes. Common methods such as the Tolman cone angle and solid angle set a universal value for each ligand regardless of the effect of steric environment on ligand geometry. We have developed two approaches to quantify sterics that improve upon the previous descriptors: the exact cone angle ($\theta°$) and the exact solid angle ($\Theta°$). The exact cone angle encapsulates a ligand inside of the smallest possible right circular cone without making any approximations concerning metal–ligand bond length or ligand geometry. The exact solid angle quantifies the area screened by a ligand on the surface of a surrounding sphere centred at the metal atom. While the cone angle only describes a single ligand, the solid angle accounts for the sterics of an entire complex. Herein, the mathematics behind each is described and numerous examples are given. An intriguing case models the change in sterics of first- and second-generation Grubbs' catalysts over the course of a ligand dissociation reaction.

Annual Reports in Computational Chemistry, Volume 9
ISSN 1574-1400
http://dx.doi.org/10.1016/B978-0-444-62672-1.00001-7

3

1. INTRODUCTION

The notion that sterics play a role in transition metal reactions began in the late 1960s with the study of metal carbonyls (1,2). Rates of substitution and exchange reactions were found to vary for different incoming nucleophiles, but generally followed a linear relationship with respect to the acid dissociation constant (pK_a) of the nucleophile. The main outlier was tricyclohexylphosphine (PCy_3), the largest nucleophile in the study, which strayed from typical behaviour. While the basicity and nucleophilicity of phosphines have been attributed to inductive effects, deviations such as PCy_3 result from the sterics of bulky substituents (3,4).

These findings opened up research into the effect of steric size on the rate constants for nucleophilic substitution (k). Steric effects were systematically separated from electronic effects by holding the pK_a of incoming nucleophiles constant while varying their physical size (5,6). Electronic profiles plotting log k versus nucleophile size suggest a "steric threshold" where once a certain bulk is reached, reaction rates drop sharply (7). This threshold is related to the conical channel through which the incoming nucleophile must travel to access the metal centre. The size of the channel provides an indication of the transition state for association: a small threshold implies a tight transition state, while a large threshold implies a loose transition state.

The sterics of ancillary ligands, which are bound to the metal centre but not directly involved in a reaction, affect the reactivity of transition metal catalysts in cross-coupling reactions (8–10). Bulky ligands accelerate reductive elimination (11) and facilitate the formation of coordinatively unsaturated complexes (12), which are often the catalytic species. Large ancillary ligands also enhance activity and regioselectivity in asymmetric cross-couplings by weakening the bonding of dissociative ligands and directing nucleophilic attack (8,13).

Ligand bulk is commonly quantified through the Tolman cone angle or the solid angle (14–16), both of which suffer from approximations that induce errors in the description. Additionally, these methods settle on a universal value for each ligand and ignore geometric changes that occur as the ligand environment evolves over the course of a reaction. Accurately quantifying steric properties of transition metal complexes in specific circumstances will lead to a better understanding of steric control and fine-tuning of catalytic species. Thus, we have introduced "exact" descriptors—in the spirit of the Tolman cone angle and the solid angle conventions—which are mathematically rigorous and contain no approximations.

2. CONE ANGLES

2.1. Cone angle history

Tolman provided the first method for quantifying steric demands of a ligand by using a cone to model symmetric phosphine ligands bound to a nickel centre *(14,17)*. The Tolman cone angle (θ) is the apex angle of a cone encapsulating the entire ligand with the edges of the cone lying on the outermost van der Waals spheres, as shown in Figure 1.1. For asymmetric phosphine ligands, the cone angle is the average of the apex angles subtended by each phosphorus substituent. Originally, the cone angle was measured using a physical space-filling model with a specialized ruler. Several empirical parameters were assumed, including a Ni–P bond distance of 2.28 Å, tetrahedral geometry about the phosphorus atom, and folding of the ligand into the smallest possible conformation. These assumptions become highly inaccurate for metals other than nickel and ligands other than monodentate phosphines. An extension was made to palladium-bound amine ligands using a Pd–N bond length of 2.20 Å, which is a rough approximation because bulky substituents on the nitrogen increase the Pd–N distance *(18)*.

In addition to geometric errors caused by these assumptions, the steric encumbrance of a ligand was assumed to be static in different environments. Phosphite ligands, having the formula $P(OR)_3$, contradict this assumption, as they are especially flexible and have been termed "steric chameleons" due to their adaptive nature *(19)*. Instead of a single preferred geometry as the

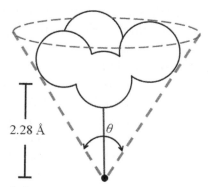

Figure 1.1 The encapsulating cone used to define the Tolman cone angle (θ) for a symmetric, monodentate phosphine ligand.

Tolman cone angle suggests, these ligands have several conformational minima observed in crystallographic structures *(20)*. Less flexible ligands also change conformation to a more compact structure in crowded environments to relieve steric strain; for example, torsion angles for triphenylphosphine (PPh$_3$) can vary up to 40° *(21)*. The central metal also affects the cone angle, as shown in a study of PPh$_3$, where increasing the period of the metal also increased the cone angle *(22)*.

To get away from using idealized physical models to approximate the cone angle, Müller *et al. (22)* developed a method to measure the Tolman cone angle for several monodentate phosphine ligands from crystallographic parameters. In this case, the encapsulating cone is defined by hydrogen atoms on the phosphorus substituents, which were assumed to have a van der Waals radius of 1.00 Å. The metal–phosphorus bond was set to 2.28 Å to match Tolman's assumption. The half apex angle for each substituent was geometrically determined from the outermost M–H bond and the M–P bond, and—analogous to the Tolman method for asymmetric ligands—the average was taken to be the cone angle. Cone angles found this way have a large range with even the simplest symmetric phosphine, PMe$_3$, which exhibits a 25.4° spread.

More recently, the Dixon and Shaughnessy groups have calculated cone angles from optimized DFT structures for zerovalent palladium complexes to determine steric effects of ancillary ligands on cross-coupling reactions *(23–26)*. For bulky ligands with cone angles over 180°, a correlation was seen between increasing size and increasing catalytic activity in Suzuki coupling reactions *(23)*. From this, the cone angle was determined to be a good measure for determining catalytic activity. In another study *(25)* on sulphonated ligands, the addition of a H$^+$ or Na$^+$ counterion to the sulphonate group enhanced the cone angle by 10° which, once again, shows that environment is an important factor in determining ligand geometry.

Due to the adaptive nature of ligands in various environments, a method for easily computing the cone angle in any situation is needed. We have found an analytical solution to the cone angle problem that relies only on the Cartesian coordinates of the ligand and van der Waals radii of ligand atoms *(27)*. Both computational and experimental researchers can benefit since geometric coordinates can be obtained by electronic structure codes as well as X-ray crystallography. The procedure sets no artificial bond lengths and assumes no idealized geometry, and thus we termed the result an exact cone angle ($\theta°$). The designation "exact" refers to the mathematical solution and not necessarily to the accuracy of the input structure. Since no empirical assumptions are made, the method is applicable to any type of ligand in any conformation bound to any metal centre. We have developed a *Mathematica (28)* program called FindConeAngle that calculates the exact cone angle and visualizes solutions for an optimized geometry. This program is freely available at www.ccqc.uga.edu.

2.2. Exact cone angle formulation

The exact cone angle ($\theta°$) is found by encapsulating the van der Waals surface of the ligand in the most acute right, circular cone that originates at the metal centre, without assuming the central metal–donor bond as the cone axis. The half apex angle α appears in the standard equation for a cone, with the apex at x_0 and axis along the unit vector n, determined by a group of tangent points x:

$$\frac{(x - x_0) \cdot n}{|x - x_0|} = \pm \cos \alpha \tag{1.1}$$

The exact cone angle comes from doubling the half-cone angle: $\theta° = 2\alpha$. A thorough derivation of the analytical solution to the exact cone angle is found in our previous report *(27)*.

The encapsulating cone is defined by one, two, or three atoms depending on the ligand geometry. In the case of a single defining atom (k), the cone angle only depends on the cone axis (n) and outer edge of the defining atom (x_k). The angle between these two vectors is the half-cone angle. An example is shown in Figure 1.2, where the nitrogen atom of an isocyanate ligand shields the carbon and oxygen atoms from the metal centre, and thus is the only determining atom.

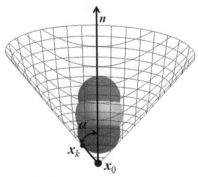

Figure 1.2 The exact cone angle ($\theta° = 2\alpha$) of an isocyanate ligand bound to a palladium centre.

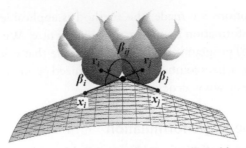

Figure 1.3 The exact cone angle of an acetylacetonate ligand bound to a chromium centre.

In the case of two defining atoms (i, j), the encapsulating cone depends on the tangent points x_i and x_j, as shown in Figure 1.3 by the ligand acetylacetonate (acac). The exact cone angle is found by the sum of three angles, each depending on the location of the atom centre and corresponding tangent point,

$$\theta° = \beta_{ij} + \beta_i + \beta_j \tag{1.2}$$

where β_{ij} is the angle between the two atom centres r_i and r_j, while (β_i, β_j) is the angle between (r_i, r_j) and the corresponding tangent point (x_i, x_j).

For the complicated case of three defining atoms (i, j, k), a key quadratic equation determines the variable $w = \cos 2\alpha$,

$$
\begin{aligned}
w^2 \left[(A-B)^2 + 4C^2 \right] + 2(A-B)(A+B-2D)w \\
+ (A+B-2D)^2 - 4C^2 = 0
\end{aligned}
\tag{1.3}
$$

where the coefficients $(A, B, C) = (u^{\mathrm{T}}Pu, v^{\mathrm{T}}Pv, u^{\mathrm{T}}Pv)$ and $D = 1 - \cos^2 \beta_{ij} - \cos^2 \beta_{jk} - \cos^2 \beta_{ik} + 2\cos \beta_{ik}\cos \beta_{ij}\cos \beta_{ij}$. These coefficients involve the vectors $u = (\cos \beta_i, \cos \beta_j, \cos \beta_k)$ and $v = (\sin \beta_i, \sin \beta_j, \sin \beta_k)$ and angles β_i and β_{ij}, which have the same definitions as in the two-atom fit. The matrix P is given by

$$
P = \begin{pmatrix}
\sin^2 \beta_{jk} & \cos \beta_{jk}\cos \beta_{ik} - \cos \beta_{ij} & \cos \beta_{ij}\cos \beta_{jk} - \cos \beta_{ik} \\
\cos \beta_{jk}\cos \beta_{ik} - \cos \beta_{ij} & \sin^2 \beta_{ik} & \cos \beta_{ik}\cos \beta_{ij} - \cos \beta_{jk} \\
\cos \beta_{ij}\cos \beta_{jk} - \cos \beta_{ik} & \cos \beta_{ik}\cos \beta_{ij} - \cos \beta_{jk} & \sin^2 \beta_{ij}
\end{pmatrix}
\tag{1.4}
$$

Solving for w gives four solutions to the cone angle: $\theta° = (\cos^{-1} w_<, 2\pi - \cos^{-1} w_<, \cos^{-1} w_>, 2\pi - \cos^{-1} w_>)$, where $w_<$ and $w_>$ denote the smaller and larger roots of the quadratic equation. Of these four, the two distinct, physical solutions that exist can be found by ascertaining whether the cone angle satisfies the equation

$$A\cos^2\alpha + B\sin^2\alpha + 2C\sin^2\alpha\cos^2\alpha = D \qquad (1.5)$$

The two remaining solutions $\theta^\circ_<$ and $\theta^\circ_>$ place the tangent spheres inside and outside of the encapsulating cone, which makes the larger root $(\theta^\circ_>)$ the desired solution.

The FindConeAngle program does a stepwise search for the cone angle by going through one-atom, two-atom, and three-atom fits in order. First, for each atom the cone axis ($\boldsymbol{n}$) is set to the vector ($\boldsymbol{m}_k$) running from the metal centre ($\mathbf{x}_0$) to atom centre ($\boldsymbol{r}_k$), and the corresponding α_k is found. If all ligand atoms lie inside of the cone created by a one-atom fit, then $\theta^\circ = 2\alpha_k$. If no one-atom fits are found, and all pairs (i, j) are iterated over to seek a two-atom fit. If all atoms lie inside of any two-atom fit, θ° results from Equation (1.2). If no two-atom fits are found, all (i, j, k) possibilities are iterated over for three-atom fits. When multiple three-atom cone solutions are found, the minimum cone angle is the exact cone angle.

2.3. Exact cone angle computations

Table 1.1 gives exact cone angles for a set of 38 monodentate ligands bound to group 10 metals. Ligand geometries were optimized using density functional theory, employing the B3LYP functional (29,30) with a 6-31G* basis set (31,32) for all first and second row atoms, and the Los Alamos effective core potential, LANL2DZ, and matching basis set for transition metals and halogens (33). Standard van der Waals radii were used in computing the cone angles (34). Geometric optimization of the ligands by five alternative functionals gave a mean difference of less than 1° in the calculated cone angles (27).

Flexible ligands were found to have several energetically similar conformations. Each phosphorous substituent can be folded up or down, as shown in Figure 1.4, to give a minimum and maximum conformation. Only ligands in which all substituents are in the "min" or "max" conformation are presented in Table 1.1. The max cone angle is on average 23° greater than the min value, though the standard deviation is quite large at 20°. In the extraordinary case of PdNBz$_3$, the max cone angle is 88.9° larger than the min value. Both max and min values should be considered to assess the steric demands of a ligand. To relieve steric crowding in a dense environment, a ligand will fold to the min conformation (35), while in an uncrowded

Table 1.1 Exact cone angles ($\theta°$) *(27)* versus Tolman cone angles (θ) *(15,17)* in degrees for monodentate ligands bound to group 10 metals

Ligand	θ	$\theta°$ (Pd)		$\theta°$ (Ni)		$\theta°$ (Pt)	
		Min	Max	Min	Max	Min	Max
PMe$_3$	118	120.4	–	125.8	–	124.7	–
PEt$_3$	132	136.0	168.7	142.6	175.8	140.4	171.1
P(n-Bu)$_3$	132	136.2	169.3	142.8	176.1	140.7	171.1
P(i-Bu)$_3$	143	–[a]	213.8	–[a]	223.7	–[a]	215.6
P(i-Pr)$_3$	160	169.0	177.1	176.6	184.7	173.2	181.4
P(t-Bu)$_3$	182	187.6	–	196.3	–	191.8	–
P(OMe)$_3$	107	111.4	170.0	129.8	176.3	118.4	173.6
P(OEt)$_3$	109	128.0	170.1	132.1	176.6	132.0	173.6
PPh$_3$	145	170.0	–	177.6	–	174.6	–
PBz$_3$	165	140.8	202.9	148.0	209.3	145.8	199.3
P(p-ClPh)$_3$	145	170.2	–	177.7	–	174.6	–
P(m-ClPh)$_3$	145	169.6	172.1	177.3	177.8	174.1	174.5
P(p-MePh)$_3$	145	170.7	–	178.2	–	174.6	–
P(m-MePh)$_3$	145	169.7	178.7	177.2	183.2	172.5	180.9
P(o-MePh)$_3$	194	175.6	208.2	184.4	218.7	179.9	210.7
P(p-OMePh)$_3$	145	170.3	171.5	177.9	178.9	173.7	175.4
P(m-(t-Bu)Ph)$_3$	145	168.4	207.3	175.9	212.6	173.3	209.8
PMe$_2$Et	123	123.7	142.3	129.3	148.7	127.8	146.8
PMe$_2$(i-Pr)	132	147.6	–	154.0	–	150.9	–
PMe$_2$(t-Bu)	139	146.0	–	152.6	–	150.2	–
PMeEt$_2$	127	129.5	153.3	135.5	160.1	133.8	156.6
PEt$_2$(i-Pr)	141	150.9	171.5	157.8	178.7	155.2	175.4
PEt$_2$(t-Bu)	149	152.7	175.7	160.1	180.7	157.1	176.8
PMe(i-Pr)$_2$	146	159.2	160.9	166.3	167.9	157.4	165.0
PEt(i-Pr)$_2$	151	169.8	174.5	177.6	182.4	173.7	179.4
P(i-Pr)$_2$(t-Bu)	167	177.2	186.6	185.3	195.3	182.4	182.9

Table 1.1 Exact cone angles ($\theta°$) *(27)* versus Tolman cone angles (θ) *(15,17)* in degrees for monodentate ligands bound to group 10 metals—cont'd

Ligand	θ	$\theta°$ (Pd) Min	Max	$\theta°$ (Ni) Min	Max	$\theta°$ (Pt) Min	Max
PMe(t-Bu)$_2$	161	174.4	–	182.2	–	179.0	–
PEt(t-Bu)$_2$	165	175.4	177.2	183.8	184.6	179.8	181.5
P(i-Pr)(t-Bu)$_2$	175	181.8	182.5	190.2	190.6	185.6	186.1
NH$_3$	94	95.4	–	112.0	–	101.9	–
NMe$_3$	132	129.6	–	140.6	–	135.4	–
NEt$_3$	150	143.9	188.1	158.8	188.6	152.1	180.8
N(i-Pr)$_3$	220	183.5	195.0	196.8	202.5	187.5	191.7
NBz$_3$	–	148.7	237.6	165.4	228.0	157.9	218.0
AsMe$_3$	114	115.5	–	120.6	–	120.6	–
AsEt$_3$	128	130.9	161.9	137.2	168.2	135.8	166.4
As(OEt)$_3$	105	122.7	161.0	126.5	167.4	125.4	166.4
AsPh$_3$	141	164.0	–	171.9	–	169.8	–

[a]Expected conformer is not a minimum on the potential energy surface.

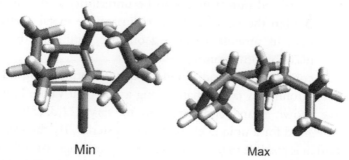

Min   Max

Figure 1.4 Idealized minimum and maximum conformations of triisobutylphosphine. The minimum conformation is not a valid structure on the potential energy surface.

environment, the ligand can relax into the max conformation. In many cases, substituents are often a mixture of min and max *(36–38)*, but when the rotational barrier of the metal–donor bond is low, the extended ligand sweeps through the rotation and the max cone angle may be the more informative value.

For the phosphine ligands listed in Table 1.1, the average M–P bond lengths for Pd, Ni, and Pt complexes were computed as 2.23, 2.05, and 2.15 Å, which are shorter than the 2.28 Å length used in the Tolman cone angle approximation. The cone angle increases as the ligand moves closer to the central metal, and only two of the phosphine ligands surveyed [P(o-MePh)$_3$, PBz$_3$] had exact cone angles smaller than the Tolman values. In general, Tolman cone angles underestimate the exact cone angle. Mean absolute deviations of the min cone angle from the Tolman value for Pd, Ni, and Pt complexes were 13.5°, 19.9°, and 16.9°. Interestingly, the deviations for nickel-bound ligands are the largest even though the Tolman cone angles were calibrated for nickel complexes. For amine ligands, the computed M–N bond distances ranged between 2.13 and 2.37 Å, whereas the Tolman method sets an approximate bond length of 2.20 Å *(18)*. The mean absolute deviation of the min cone angle from the Tolman value for Pd, Ni, and Pt complexes is 11.6°, 14.7°, and 11.5°, respectively.

For triisobutylphosphine, P(i-Bu)$_3$, the Tolman cone angle (143°) has more than a 70° difference compared with the exact cone angle for the Ni-bound ligand (223.7°). This disparity is due to the idealized space-filling model used in the evaluation of Tolman cone angles and the assumption of the most compact structure. In this min structure, each CH(CH$_3$)$_2$ is facing up, while the CH$_2$ groups point down toward the metal centre (Figure 1.4). We found the idealized min structure to be unrealistic, as the methylene hydrogens are within the van der Waals radius of the metal centre. Our B3LYP computations provide a single geometric minimum that corresponds to the max conformation.

The exact cone angle method allows for quantification of ligand sterics for specific environments rather than setting a universal value. Numerous reports have shown ligand conformations to vary *(21,35,39,40)*, propounding the need for a unique classification of sterics. By the exact cone angle method, it is possible to track the change in cone angle over the course of a reaction to evaluate steric control and the role of ancillary ligand size in organometallic reactions.

3. SOLID ANGLES

3.1. Solid angle history

An alternative measure of steric demand is the solid cone angle (Θ), derived from the solid angle (Ω) which is the area of the shadow cast by ligand atoms onto a sphere centred about the central metal atom *(16)*. Figure 1.5 displays

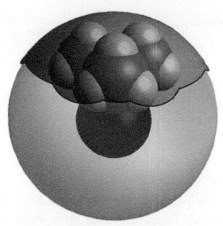

Figure 1.5 The solid angle (Ω) of palladium trimethylphosphine (dark) defined by the shadow cast by ligand atoms onto a hypothetical sphere.

the shadow cast by a trimethylphosphine (PMe$_3$) ligand upon illumination from the palladium centre. The general equation for the solid angle of a shadow (S) on the surface is given by the integral $\Omega = \int_S \frac{r \cdot d\sigma}{r^3}$, where r is the radius vector of the sphere, r is the vector length, and $d\sigma$ is the vector element of area at the surface *(41)*. The solid angle for a single sphere i can be calculated analytically by the equation $\Omega_i = 2\pi[1 - \cos(\alpha_i)]$, where α_i is the semi-vertex angle of the shadow cone *(42)*. In the case of multiple, non-overlapping shadows, the total solid angle is the sum of all one-atom solid angles. Non-overlapping shadows, however, rarely occur and often multiple overlap regions are present.

In one method for determining Ω, the solid angle for each atom is summed and the overlap regions are subtracted off to give the total solid angle. In cases of multiple overlap, the shadow regions are projected onto an underlying plane—adding an approximation to Ω—and complicated subtraction schemes are performed to remove the overlap area from the total Ω *(42)*. In a single CH$_3$ fragment, there are regions of quadruple overlap, and a four-step procedure is needed to correct for the total solid angle. Regions of more complicated overlap can be simplified by dividing the total ligand into simpler fragments to give a semi-quantitative solid angle *(43)*.

Instead of individually summing shadows and finding the area of overlap, we have developed an analytic solution to determine the collective solid angle *(44)*. In this approach, a perimeter of the entire ligand shadow is found and a line integral is computed to give the exact solid angle ($\Omega°$). For comparison to the Tolman cone angle, the exact solid cone angle ($\Theta°$) is then found through the equation $\Theta^{\circ} = 2cos^{-1}\left(1 - {\Omega^{\circ}}/{2\pi}\right)$. In keeping with previous work, we use the word "exact" to describe the mathematics and not the ligand geometry itself. As with the exact cone angle, all that is needed is the van der Waals radii and Cartesian coordinates of ligand atoms. A *Mathematica (28)* program called FindSolidAngle, which calculates the exact solid angle and visualizes solutions, is freely available at www.ccqc. uga.edu.

3.2. Exact solid angle formulation

To compute Ω° via a line integral, the perimeter of the shadow region must be determined. For each ligand atom, the apex angle β_i and cone axis $\mathbf{r}_i$, shown in Figure 1.3, define a shadow cone that is projected onto a unit sphere. Intersecting arcs formed by the boundaries of these shadows make up the perimeter, as illustrated in Figure 1.6. For each boundary arc, the change in azimuthal angle ($\Delta\phi$) and the turn angle (τ) of each intersection must be known. Each perimeter comprises a single loop, and the contributions from each loop are added together to find the total solid angle. A detailed derivation for finding the azimuthal and turn angles is given in our forthcoming publication *(44)*.

Two types of loops, convex and concave, can exist. Convex loops are formed around the outer perimeter of the total shadow projection. Concave loops are formed by gaps in overlapping shadows and are subtracted from the

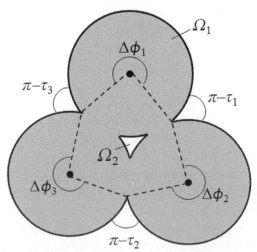

Figure 1.6 The total solid angle of a convex loop Ω_1 enclosing a concave loop Ω_2.

enclosing convex loop to evaluate the total solid angle. The line integrals for each loop l that yield the total solid angle simplify greatly into an exact equation for the total solid angle,

$$\Omega^{\circ} = \sum_{l}^{\text{loops}} \left(2\pi + \sum_{k}^{n} \tau_{k,l} - \sum_{k}^{\text{arcs}} |\Delta\phi_{k,l}| \cos\alpha_{k,l} \right) - 4\pi N_{\text{concave}} \quad (1.6)$$

where k runs over all boundary arcs, $\alpha_{k,l}$ is the half apex angle of shadow cone k in loop l, and N_{concave} is the number of concave loops contained within an outer convex loop.

The exact solid angle procedure has been implemented in the FindSolidAngle program, and the user only needs to input the optimized or experimentally found Cartesian coordinates of the complex to determine Ω°. Like the exact cone angle, the exact solid angle allows mapping of changes in the steric profile of a complex over the course of a reaction.

3.3. Exact solid angle computations

The solid angle provides an instantaneous snapshot of the ligand that is not modelled in the cone angle description, which is designed for environments with free rotation. Rigid molecules, therefore, are well described by the solid angle. Additionally, the solid angle can be calculated for an entire complex rather than a single ligand to quantify the total steric shielding of a metal centre *(45,46)*. However, as solid angles are extremely sensitive to the geometry of the ligand *(20)*, they must be computed anew for each complex.

Table 1.2 shows exact solid cone angles (Θ°) computed for numerous monodentate phosphine, amine, and arsenic ligands bound to Pd, Ni, and Pt. Due to gaps in the ligand shadow, the exact solid cone angles are systematically smaller than the exact cone angles. Although solid angles are additive, solid cone angles do not increase in any linear fashion because of overlap effects. The solid cone angles for min conformers of Pd-bound triethylphosphine (PEt$_3$) and tributylphosphine (P(n–Bu)$_3$) are extremely similar (119.0° and 119.1°) despite P(n–Bu)$_3$ having a carbon chain twice as long. The depictions of the solid angle for the two cases look identical (Figure 1.7). When nickel is the metal centre, PEt$_3$ actually has a slightly larger solid cone angle than P(n–Bu)$_3$ (128.1° vs. 127.9°).

The barrier to rotation about the metal–ligand bonds increases in cases of intramolecular repulsion between ligands *(47)*. Since the solid angle allows gaps in the ligand profile, regions for meshing with adjacent ligands can be viewed with solid angle plots. To recognize the effect of gaps on the steric

Table 1.2 Computed exact solid cone angles ($\Theta°$) *(44)* versus reported solid cone angles (Θ) *(15)* in degrees for monodentate ligands bound to group 10 metals

Ligand	Θ	$\Theta°$ (Pd) Min	$\Theta°$ (Pd) Max	$\Theta°$ (Ni) Min	$\Theta°$ (Ni) Max	$\Theta°$ (Pt) Min	$\Theta°$ (Pt) Max
PMe$_3$	124	114.3	–	124.3	–	119.2	–
PEt$_3$	143	119.0	133.6	128.1	142.0	123.6	136.7
P(n–Bu)$_3$	148	119.1	134.6	127.9	143.0	123.5	138.2
P(i–Bu)$_3$	173	–[a]	174.5	–[a]	184.5	–[a]	176.9
P(i–Pr)$_3$	163	138.7	150.4	146.2	156.7	142.9	154.2
P(t–Bu)$_3$	182	160.8	–	168.3	–	164.7	–
P(OMe)$_3$	113	111.4	134.2	129.8	144.8	118.4	138.5
P(OEt)$_3$	117	114.5	137.0	129.7	146.2	120.6	140.1
PPh$_3$	129	135.7	–	146.0	–	140.6	–
PBz$_3$	163	123.0	159.3	131.5	167.7	127.5	158.9
P(p–ClPh)$_3$	129	135.8	–	146.2	–	140.7	–
P(m–ClPh)$_3$	136	135.6	137.3	146.2	147.3	140.5	141.7
P(p–MePh)$_3$	135	135.9	–	146.0	–	140.4	–
P(m–MePh)$_3$	140	135.7	138.8	145.8	148.1	139.7	142.7
P(o–MePh)$_3$	142	141.9	159.9	151.7	169.8	146.4	162.0
P(p–OMePh)$_3$	139	135.8	136.1	145.9	146.2	140.2	140.7
P(m–(t–Bu)Ph)$_3$	159	135.4	154.2	145.3	162.8	140.0	157.7
PMe$_2$Et	133	115.7	120.7	125.1	129.7	120.4	125.3
PMe$_2$(i–Pr)	140	122.9	–	131.5	–	127.0	–
PMe$_2$(t–Bu)	144	129.0	–	136.9	–	133.0	–
PMeEt$_2$	138	117.3	126.8	126.4	135.5	122.0	130.7
PEt$_2$(i–Pr)	150	125.5	140.6	134.0	148.8	130.0	144.6
PEt$_2$(t–Bu)	156	133.2	143.8	141.3	148.9	137.4	145.0
PMe(i–Pr)$_2$	151	134.2	137.8	142.8	145.3	134.2	141.6
PEt(i–Pr)$_2$	156	137.5	145.7	146.5	153.2	141.5	149.9
P(i–Pr)$_2$(t–Bu)	170	152.0	153.2	160.3	160.3	151.2	157.0

Table 1.2 Computed exact solid cone angles ($\Theta°$) (44) versus reported solid cone angles (Θ) (15) in degrees for monodentate ligands bound to group 10 metals—cont'd

Ligand	Θ	$\Theta°$ (Pd)		$\Theta°$ (Ni)		$\Theta°$ (Pt)	
		Min	Max	Min	Max	Min	Max
PMe(t-Bu)$_2$	163	144.7	–	152.2	–	148.8	–
PEt(t-Bu)$_2$	161	147.8	150.5	155.2	157.5	151.7	154.3
P(i-Pr)(t-Bu)$_2$	178	154.3	156.1	161.8	163.3	158.3	160.1
NH$_3$	87	93.8	–	112.0	–	101.9	–
NMe$_3$	124	120.5	–	131.8	–	126.4	–
NEt$_3$	142	125.9	148.3	138.6	154.3	132.3	148.0
N(i-Pr)$_3$	161	152.1	166.1	165.6	176.4	157.7	168.5
NBz$_3$	173	129.9	156.1	143.2	181.2	137.2	170.6
AsMe$_3$	126	109.7	–	118.5	–	114.9	–
AsEt$_3$	134	114.0	128.0	122.7	135.3	118.9	132.3
As(OEt)$_3$	117	109.7	129.9	123.3	138.8	114.8	134.3
AsPh$_3$	128	130.0	–	139.7	–	135.2	–

[a]Expected conformer is not a minimum on the potential energy surface.

profile, consider the exact descriptors of PdPPh$_3$. The exact cone angle is $\theta° = 170.0°$, while the exact solid cone angle for the same geometry is $\Theta° = 135.7°$. As shown in Figure 1.8, the solid angle of PPh$_3$ retains C_3-like symmetry with gaps between each phenyl substituent. These gaps amount to 20% of the exact cone angle.

Multidentate ligands, which are rotationally hindered, are well described by the solid angle. Bidentate phosphines, used as ancillary ligands in cross-coupling reactions (48,49), often consist of a carbon chain backbone with phosphorus substituents of varying size. The ligand 1,2-bis(-diphenylphosphino)ethane (dppe) has a two-carbon backbone and phenyl substituents on the phosphorus atoms, as depicted in Figure 1.9. The exact solid cone angle is 175.0°. The thin backbone causes indentations in the solid angle not accounted for in the exact cone angle method, which gives a much larger value of $\theta° = 218.6°$ (27). Since there is no rotation about the M–P bond, $\theta°$ may overestimate the steric encumbrance. The exact solid cone angle, along with the solid angle plot, gives a more detailed description of the ligand sterics.

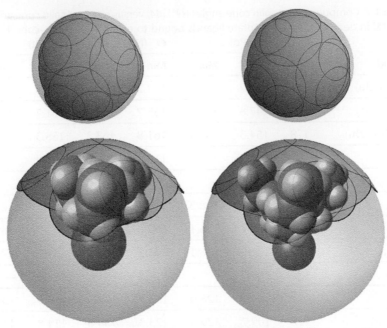

Figure 1.7 Solid angles of the min conformations of Pd(PEt₃) (left) and Pd(P(n-Bu)₃) (right) appear virtually identical from the top and side views.

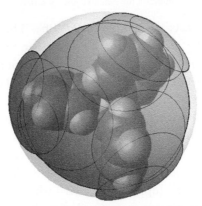

Figure 1.8 The solid angle of palladium-bound triphenylphosphine (PPh₃) is represented by the shaded area.

Previously, empirical extensions of the original solid angle calculations were made based on the bite angle and metal–donor bond lengths *(50)*. The resultant parameter, called the generalized equivalent cone angle (Θ_b), was determined for a set of crystallographic structures for bidentate phosphine ligands bound to platinum centres. Many crystallographic

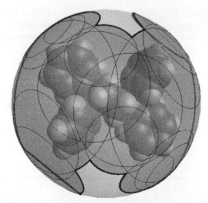

Figure 1.9 The solid angle of Pd(1,2-bis(diphenylphosphino)ethane) shows gaps in the steric profile.

structures—all containing different neighbouring ligands—were analyzed to give average Θ_b parameters, with deviations ranging from 1.3° to 3.3°. Table 1.3 contains a comparison of this empirical descriptor to the exact solid cone angle for both platinum- and palladium-bound ligands geometrically optimized in the absence of other ligands. The mean absolute deviation from Θ_b for platinum structures is 6.1° with a standard deviation of 5.7°. Only one ligand (dppm) lies within the experimental standard deviation. The mean absolute deviation for palladium structures is 4.0° also with a standard deviation of 5.7°. In this case, five ligands (dmpe, d(i-Pr)pe, d(t-Bu)pp, dcpe, and dppp) lie within the experimental standard deviation. However, since Θ_b values were found for complexes containing neighbouring ligands, it is ambiguous as to whether the empirical correction or geometric distortions caused the disparity. Nevertheless, the exact solid cone angle has the advantage of quantifying the steric size of a ligand by computational analysis rather than crystallization, which is often difficult and time consuming.

Ligands with large holes are better quantified by the solid angle than the cone angle. For example, in K^+(18-crown-6) the crown ether forms a belt around the central potassium atom. The total solid cone angle, depicted on the left in Figure 1.10, is 197°, and each hole is 110°. Incoming species smaller than 110° could potentially access the potassium centre through the hole. The exact cone angle cannot account for the entirety of the ligand, though it can quantify the size of the conical channel through the hole, shown on the right in Figure 1.10. This channel was found to be 101.2°, slightly smaller than the solid angle estimate.

Table 1.3 Experimental (Θ_b) (50) and exact (Θ°) (44) solid cone angles in degrees

Ligand	Θ_b	Θ° (Pt)	Θ° (Pd)
dmpe	155.5 ± 1.6	161.4	156.1
depe	175.5 ± 4.2	168.6	163.7
d(i-Pr)pe	187.6 ± 1.7	192.3	187.4
d(i-Pr)pp	192.5 ± 1.3	200.9	195.1
d(t-Bu)pe	202.6 ± 1.4	204.7	200.0
d(t-Bu)pp	211.0 ± 1.3	215.9	211.3
dcpe	191.0 ± 2.4	193.6	188.9
dppm	168.1 ± 3.3	167.4	160.0
dppe	177.6 ± 2.5	180.7	175.0
dppp	182.9 ± 2.9	190.4	184.2
dppb	187.7 ± 2.5	201.8	195.9
dppf	192.2 ± 2.2	204.8	199.8

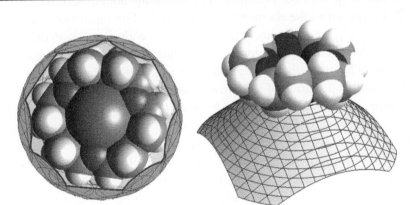

Figure 1.10 K^+(18-crown-6) described by the solid angle (left) and cone angle (right).

Unlike the cone angle, the solid angle can simultaneously describe the sterics of any number of bound ligands. Examining the crowding of multiple ligands can lead to the study of sterics for entire catalytic complexes. For example, the Grubbs' second-generation catalyst is more stable and active than the first-generation catalyst (51–53). The only difference between these two species is the substitution of a phosphine ligand by an N-heterocyclic carbene ligand. Dissociation of the remaining phosphine ligand (PCy_3) generates the active 14-electron species. Contrary to reactivity of the

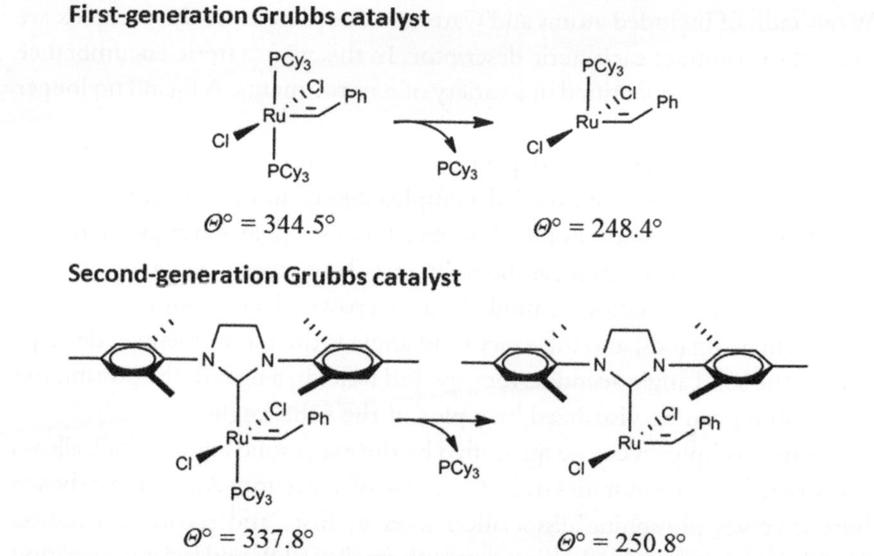

First-generation Grubbs catalyst

$\Theta° = 344.5°$ $\Theta° = 248.4°$

Second-generation Grubbs catalyst

$\Theta° = 337.8°$ $\Theta° = 250.8°$

Figure 1.11 Exact solid cone angles ($\Theta°$) of the entire Grubbs catalyst—including all ligands—before and after phosphine dissociation. *Geometries from ref. (56)*.

complexes, phosphine dissociation for the first-generation catalyst is faster than the second-generation catalyst *(54,55)*. A study on computed bond dissociation energies (BDEs) confirmed a higher BDE for the second-generation catalyst *(56)*. By examining the exact solid angle for each complex, we can ascribe this surprising trend to sterics. The first-generation catalyst, $(PCy_3)_2Cl_2Ru=CHPh$, has a solid angle $6.7°$ larger than the second-generation catalyst, $(H_2IMes)(PCy_3)Cl_2Ru=CHPh$ (Figure 1.11). Upon PCy_3 dissociation, the solid angles decrease by $96.1°$ and $87.0°$, respectively. The first-generation catalyst is more crowded before dissociation and less crowded after dissociation. Losing the PCy_3 ligand reduces steric interactions more for the first-generation than for the second-generation catalyst. The decrease in steric interactions potentially promotes phosphine dissociation.

4. CONCLUSION

Mathematically rigorous approaches to quantify the size of ligands are needed to accurately describe the sterics of a complex. We have provided methods analogous to the Tolman cone angle and solid angle, termed the exact cone angle and exact solid angle, which, unlike their earlier counterparts, require no approximations or idealized geometries. Only the van der

Waals radii of included atoms and Cartesian coordinates of the complex are needed to compute each steric descriptor. In this way, a steric encumbrance of a ligand can be quantified in a variety of environments. A ligand no longer needs a universal value.

The choice of descriptor depends on the complex itself. A monodentate ligand in a relatively uncrowded complex likely undergoes free rotation about the metal–donor bond. The exact cone angle description is recommended in this situation, as the right, circular cone imitates a freely rotating ligand. Free rotation is unlikely in a crowded environment or with multidentate ligands, and the exact solid angle is the recommended descriptor. As the solid angle accounts for gaps and holes in a ligand, the possibilities for meshing can be visualized by a plot of the solid angle.

Entire complexes can be quantified by the exact solid angle, which allows study of the change in sterics over the course of a reaction. An example shown here involves phosphine dissociation rates in first- and second-generation Grubbs' catalysts, where the complex with smaller BDE and faster dissociation rate had a larger solid angle before dissociation and smaller solid angle after dissociation. The steric crowding possibly promotes phosphine dissociation, though further studies would be needed to confirm this conjecture.

REFERENCES

1. Schuster-Woldan, H. G.; Basolo, F. *J. Am. Chem. Soc.* **1966**, *88*, 1657.
2. Thorsteinson, E. M.; Basolo, F. *J. Am. Chem. Soc.* **1966**, *88*, 3929.
3. Henderson, W. A.; Streuli, C. A. *J. Am. Chem. Soc.* **1960**, *82*, 5791.
4. Henderson, W. A.; Buckler, S. A. *J. Am. Chem. Soc.* **1960**, *82*, 5794.
5. Schenkluhn, H.; Scheidt, W.; Weimann, B.; Zähres, M. *Angew. Chem. Int. Ed. Engl.* **1979**, *18*, 401.
6. Zizelman, P. M.; Amatore, C.; Kochi, J. K. *J. Am. Chem. Soc.* **1984**, *106*, 3771.
7. Golovin, M. N.; Rahman, M. M.; Belmonte, J. E.; Giering, W. P. *Organometallics* **1985**, *4*, 1981.
8. Kranenburg, M.; van der Burgt, Y. E. M.; Kamer, P. C. J.; van Leeuwen, P. W. N. M.; Goubitz, K.; Fraanje, J. *Organometallics* **1995**, *14*, 3081.
9. Freixa, Z.; van Leeuwen, P. W. N. M. *Dalton Trans.* **2003**, 1890.
10. Dierkes, P.; van Leeuwen, P. W. N. M. *J. Chem. Soc., Dalton Trans.* **1999**, 1519.
11. Hartwig, J. F.; Richards, S.; Barañano, D.; Paul, F. *J. Am. Chem. Soc.* **1996**, *118*, 3626.
12. Hermes, A. R.; Girolami, G. S. *Inorg. Chem.* **1988**, *27*, 1775.
13. van Haaren, R. J.; Goubitz, K.; Fraanje, J.; van Strijdonck, G. P. F.; Oevering, H.; Coussens, B.; Reek, J. N. H.; Kamer, P. C. J.; van Leeuwen, P. W. N. M. *Inorg. Chem.* **2001**, *40*, 3363.
14. Tolman, C. A.; Seidel, W. C.; Gosser, L. W. *J. Am. Chem. Soc.* **1974**, *96*, 53.
15. Brown, T. L.; Lee, K. *J. Coord. Chem. Rev.* **1993**, *128*, 89.
16. Immirzi, A.; Musco, A. *Inorg. Chim. Acta* **1977**, *25*, L41.
17. Tolman, C. A. *Chem. Rev.* **1977**, 77, 313.
18. Seligson, A. L.; Trogler, W. C. *J. Am. Chem. Soc.* **1991**, *113*, 2520.
19. Smith, J. M.; Coville, N. *J. Organomet.* **2001**, *20*, 1210.

20. Smith, J. M.; Coville, N. J.; Cook, L. M.; Boeyens, J. C. A. *Organometallics* **2000**, *19*, 5273.
21. Martín, A.; Orpen, A. G. *J. Am. Chem. Soc.* **1996**, *118*, 1464.
22. Müller, T. E.; Mingos, D. M. P. *Transition Metal Chem.* **1995**, *20*, 533.
23. DeVasher, R. B.; Spruell, J. M.; Dixon, D. A.; Broker, G. A.; Griffin, S. T.; Rogers, R. D.; Shaughnessy, K. H. *Organometallics* **2005**, *24*, 962.
24. Hill, L. L.; Moore, L. R.; Huang, R.; Craciun, R.; Vincent, A. J.; Dixon, D. A.; Chou, J.; Woltermann, C. J.; Shaughnessy, K. H. *J. Org. Chem.* **2006**, *71*, 5117.
25. Moore, L. R.; Western, E. C.; Craciun, R.; Spruell, J. M.; Dixon, D. A.; O'Halloran, K. P.; Shaughnessy, K. H. *Organometallics* **2008**, *27*, 576.
26. Hill, L. L.; Smith, J. M.; Brown, W. S.; Moore, L. R.; Guevera, P.; Pair, E. S.; Porter, J.; Chou, J.; Wolterman, C. J.; Craciun, R.; Dixon, D. A.; Shaughnessy, K. H. *Tetrahedron* **2008**, *64*, 6920.
27. Bilbrey, J. A.; Kazez, A. H.; Locklin, J.; Allen, W. D. *J. Comp. Chem.* **2013**, *34*, 1189–1197. http://dx.doi.org/10.1002/jcc.23217.
28. Wolfram Research, Inc., *Mathematica 8.0*, **2010**, Champaign, IL.
29. Lee, C.; Yang, W.; Parr, R. G. *Phys. Rev. B* **1988**, *37*, 785.
30. Becke, A. D. *J. Chem. Phys.* **1993**, *98*, 5648.
31. Dill, J. D.; Pople, J. A. *J. Chem. Phys.* **1975**, *62*, 2921.
32. Francl, M. M.; Pietro, W. J.; Hehre, W. J.; Binkley, J. S.; Gordon, M. S.; DeFrees, D. J.; Pople, J. A. *J. Chem. Phys.* **1982**, 77, 3654.
33. Hay, P. J.; Wadt, W. R. *J. Chem. Phys.* **1985**, *82*, 299.
34. Bondi, A. *J. Phys. Chem.* **1964**, *68*, 441.
35. Bunten, K. A.; Chen, L.; Fernandez, A. L.; Poë, A. *J. Coord. Chem. Rev.* **2002**, *233–234*, 41.
36. Munro-Leighton, C.; Adduci, L. L.; Becker, J. J.; Gagné, M. R. *Organometallics* **2011**, *30*, 2646.
37. Charmant, J. P. H.; Espinet, P.; Soulantica, K. *Acta Crystallogr., Sect. E: Struct. Rep. Online* **2001**, *57*, m451.
38. Englert, U.; Matern, E.; Olkowska-Oetzel, J.; Pikies, J. *Acta Crystallogr., Sect. E: Struct. Rep. Online* **2003**, *59*, m376.
39. Ferguson, G.; Roberts, P. J.; Alyea, E. C.; Khan, M. *Inorg. Chem.* **1978**, *17*, 2965.
40. Bart, J. C. J.; Favini, G.; Todeschini, R. *Phosphorous Sulfur Rel. Elem.* **1983**, *17*, 205.
41. White, D.; Taverner, B. C.; Leach, P. G. L.; Coville, N. J. *J. Comp. Chem.* **1993**, *14*, 1042.
42. Taverner, B. C. *J. Comp. Chem.* **1996**, *17*, 1612.
43. White, D.; Coville, N. J. *Adv. Organomet. Chem.* **1994**, *36*, 95.
44. Bilbrey, J. A.; Kazez, A. H.; Locklin, J.; Allen, W. D. *In Preparation*. **2013**.
45. Fukin, G. K.; Guzei, I. A.; Baranov, E. V. *J. Coord. Chem.* **2007**, *60*, 937.
46. Guzei, I. A.; Sánchez-Castro, M. E.; Ramirez-Monroy, A.; Cervantes-Vásquez, M.; Alemán Figueroa, I. R.; Paz-Sandoval, M. A. *Inorg. Chim. Acta* **2006**, *359*, 701.
47. Fanizzi, F. P.; Lanfranchi, M.; Natile, G.; Tiripicchio, A. *Inorg. Chem.* **1994**, *33*, 3331.
48. Tamao, K.; Kodama, S.; Nakajima, I.; Kumada, M.; Minato, A.; Suzuki, K. *Tetrahedron* **1982**, *38*, 3347.
49. Percec, V.; Golding, G. M.; Smidrkal, J.; Weichold, O. *J. Org. Chem.* **2004**, *69*, 3447.
50. Niksch, T.; Görls, H.; Weigand, W. *Eur. J. Inorg. Chem.* **2010**, *2010*, 95.
51. Scholl, M.; Ding, S.; Lee, C. W.; Grubbs, R. H. *Org. Lett.* **1999**, *1*, 953.
52. Huang, J.; Stevens, E. D.; Nolan, S. P.; Petersen, J. L. *J. Am. Chem. Soc.* **1999**, *121*, 2674.
53. Bielawski, C. W.; Grubbs, R. H. *Angew. Chem. Int. Ed. Engl.* **2000**, *39*, 2903.
54. Sanford, M. S.; Love, J. A.; Grubbs, R. H. *J. Am. Chem. Soc.* **2001**, *123*, 6543.
55. Sanford, M. S.; Ulman, M.; Grubbs, R. H. *J. Am. Chem. Soc.* **2001**, *123*, 749.
56. Zhao, Y.; Truhlar, D. G. *Org. Lett.* **2007**, *9*, 1967.

CHAPTER TWO

Efficient Monomer-Based Quantum Chemistry Methods for Molecular and Ionic Clusters

Leif D. Jacobson*, **Ryan M. Richard**[†], **Ka Un Lao**[†], **John M. Herbert**[†,1]

*Department of Chemistry, Yale University, New Haven, Connecticut, USA
[†]Department of Chemistry and Biochemistry, The Ohio State University, Columbus, Ohio, USA
[1]Corresponding author: e-mail address: herbert@chemistry.ohio-state.edu

Contents

Abstract

Fragment-based quantum chemistry is an increasingly popular means to extend electronic structure theory to large systems. A variety of fragment-based approaches exist, and we provide an overview of several that we have developed recently. Our goal is not only to extend quantum chemistry calculations to large clusters and molecular liquids, but also to achieve "chemical" accuracy of $\lesssim 1$ kcal/mol with respect to the best available *ab initio* benchmarks. We outline several routes to achieving this level of accuracy. The first is a generalization of the traditional many-body expansion to include situations where the monomer units overlap, which provides good accuracy for clusters of polar monomers even when the many-body expansion is truncated at the two-body level. A second method generalizes the technique of symmetry-adapted perturbation theory to many-body clusters and works especially well when combined with empirical dispersion potentials. Each of these methods is embarrassingly parallelizable and thus amenable to sizable clusters.

Annual Reports in Computational Chemistry, Volume 9
ISSN 1574-1400
http://dx.doi.org/10.1016/B978-0-444-62672-1.00002-9

25

1. INTRODUCTION AND OVERVIEW

Electronic structure theory is a mature field, in which incredible advances have been made over several decades. High accuracy ground-state potential energy surfaces are available for small molecules from first-principles wave function methods, and plane-wave implementations of density functional theory (DFT) allow routine computations on solids with sizable unit cells. DFT calculations on systems containing $\sim 10^5$ atoms have been performed on large supercomputers *(1)*, and scalable parallel implementations of highly-correlated wave function methods such as CCSD(T) have been extended, for example, to $(H_2O)_{17}$ in a triple-ζ basis set (1564 basis functions), a calculation that required several hours on 120,000 processors for the triples correction alone *(2)*. Even the DFT approach is not inexpensive for ~ 1500 basis functions, however, especially where optimizations or dynamics are required. As such, there is a pronounced need for quantum chemistry models that can realistically be applied to molecular liquids and other large systems where an accurate (but computationally feasible) treatment of non–covalent interactions would be useful. Preliminary inroads in this direction are the topic of this report.

Non-covalent interactions underpin much of chemical physics and are responsible for a vast array of chemical phenomena. The van der Waals or dispersion interaction, for example, is responsible for the weak interaction between two argon atoms (which is nevertheless sufficient to bind six vibrational levels *(3)*); for the steady increase in alkane boiling points as a function of chain length; for π-stacking effects in DNA, which—contrary to popular belief—may be more important to the stability of the double helix than are hydrogen-bonding interactions *(4)*; and for the adhesive abilities of the gecko *(5)*. Competition between various types of intermolecular interactions is also important, as in the aforementioned DNA example and also in many molecular liquids and solids. For example, the interplay between parallel and perpendicular orientations of C_2H_2, which arises from competition between electrostatic and dispersion interactions, gives rise to the well-known herringbone structure of crystalline acetylene *(6)*, and neutron diffraction studies of liquid benzene reveal a similar competition between perpendicular and π-stacked orientations *(7)*.

Although central to chemistry, dispersion interactions arise purely from electron correlation effects, meaning that they are absent at the Hartree–Fock (HF) level. Dispersion interactions are also described poorly by many of the most popular density-functional approximations *(8)* and must be "grafted on" after the fact, by means of parameterized empirical dispersion potentials (DFT + D) *(9)*. Even CCSD calculations significantly underestimate dispersion interactions in certain π complexes *(10)* and represent only a modest improvement to MP2 calculations for benchmark databases of non-covalent dimers (root-mean-square errors of 0.7–1.0 kcal/mol) *(11)*. This makes the proper description of non-covalent interactions a challenging problem for computational chemistry, even more so because one would like to move beyond dimer binding energies, to explore larger clusters and molecular liquids.

One way to imagine reducing the computational cost is to partition the "supersystem" into smaller subsystems or fragments, then appeal to the many-body expansion (MBE),

$$E = \sum_{I=1}^{N} E_I + \sum_{I=1}^{N} \sum_{J<I} \Delta E_{IJ} + \sum_{I=1}^{N} \sum_{J<I} \sum_{K<J} \Delta E_{IJK} + \cdots, \qquad (2.1)$$

to approximate the total energy as a sum of monomer energies (E_I), pairwise corrections

$$\Delta E_{IJ} = E_{IJ} - E_I - E_J, \qquad (2.2)$$

three-body corrections

$$\Delta E_{IJK} = E_{IJK} - \Delta E_{IJ} - \Delta E_{JK} - \Delta E_{IK} - E_I - E_J - E_K, \qquad (2.3)$$

etc. *(12,13)*. A variety of fragment-based methods have been developed over the last 10–15 years *(12)* and applied not just to non-covalent clusters and molecular crystals but also to macromolecular systems. The notion of fragment-based quantum chemistry seems especially natural in the context of intermolecular interactions, however, and is in fact the basis of the well-established symmetry-adapted perturbation theory (SAPT) approach to computing non-covalent interactions *(14–16)*, wherein monomer wave functions serve as the zeroth-order states for a perturbative treatment of the intermolecular interaction, which is expanded directly rather than computed as an energy difference.

If the series in Equation (2.1) is carried out to include N-body terms, where N is the number of monomer units, then the expansion is exact but not useful. Its utility is that it provides a spatially homogeneous way to approximate the supersystem energy *(17)*, as opposed to more traditional QM/MM methods or QM:QM (ONIOM-type *(18)*) models, where different spatial regions are described at different levels of theory. If the MBE can be truncated at low order then progress can be made, especially if different orders can be treated at different levels of theory *(19–22)*. For systems composed of polar monomers, however, many-body polarization effects are significant and even the four-body terms may not be negligible *(17,19,23,24)*. In such cases, convergence of the MBE is greatly accelerated by embedding each monomer, dimer, trimer, etc., in an electrostatic potential representing the other monomers *(25–28)*.

Other research has focused on dividing up a large supersystem into overlapping fragments that contain some common nuclei *(29–37)*, which may be especially fruitful for macromolecular systems, where "edge effects" (due to the capping of severed valencies when fragments are drawn across covalent bonds) may be significant. Even limiting the discussion to non-covalent clusters, however, it is not obvious *a priori* that limiting the MBE to one monomer per fragment is ideal, in terms of cost per unit accuracy. For clusters composed of small, polar monomers, we find that the use of overlapping fragments with three to four monomer units per fragment provides very high accuracy even at the two-body level *(37,38)*, as detailed in Section 2.

The plethora of fragment-based methods is further expanded when one considers that a variety of electrostatic embedding schemes are possible, ranging from atom-centred point charges (or point charges plus point dipoles *(39)*) to proper monomer electron densities *(40)*. These considerations alone suggest an almost unwieldy variety of different approximation schemes, as a recent review of the literature confirms *(12)*. We have recently sought to unify (or at least classify and categorize) these seemingly disparate approaches, by means of a generalized many-body expansion (GMBE) that extends Equation (2.1) to the case of overlapping fragments but reduces to Equation (2.1), order-by-order, if the fragments are disjoint *(37,38)*. This allows us to view existing fragment-based quantum chemistry methods as different approximations to a common energy formula and suggests an elemental way in which to define a particular method. It also suggests some new fragment-based methods involving dimers of overlapping fragments, which show promise for high accuracy calculations. These developments will be discussed in Section 2.

The fact that electrostatic embedding accelerates the convergence of the MBE suggests that electronic polarization plays an important role in the higher-order terms in Equation (2.1). In the case of water clusters, a supersystem HF calculation followed by a two-body (pairwise additive) expansion of the Møller–Plesset MP2 correlation energy yields an excellent approximation to the energy obtained from a supersystem MP2 calculation (25,26,41–43). The requirement of a supersystem HF calculation makes this an expensive approach for large clusters, however, even if it is significantly cheaper than a supersystem MP2 calculation. Beran and co-workers (19–21) have eliminated this bottleneck by computing three- and four-body interactions with a polarizable force field and combining this with monomer and dimer MP2 calculations, again with excellent results as compared to supersystem MP2 calculations for clusters of polar molecules. Together, these observations suggest that although polarization (induction) has important many-body contributions, electron correlation effects are largely pairwise additive. This is consistent with the results of three-body SAPT calculations and energy decomposition analysis in small clusters, both of which suggest that dispersion and exchange-repulsion interactions are indeed pairwise-additive to a good approximation (44–49).

These observations motivated us to develop a framework to incorporate many-body induction effects in a self-consistent way, using a monomer-based self-consistent field (SCF) algorithm (50) whose cost grows only linearly with the number of monomer units, and then to incorporate the remaining contributions to the intermolecular interaction via a pairwise-additive form of SAPT (51,52). The resulting method, which we initially developed in an intuitive way based on the aforementioned considerations, can in fact be rigorously derived based on systematic approximations to the Schrödinger equation for the supersystem (52). This method will be discussed, with some illustrative examples of its successes and failures, in Section 3.

Finally, in Section 4, we describe very recent work that aims to replace the most expensive (and least accurate) terms in this theory—namely, the dispersion and exchange-dispersion terms—with empirical atom–atom potentials, similar to DFT + D approaches (53,54). The result is a quantum chemistry method for molecular and ionic clusters that is high in accuracy (mean error <0.4 kcal/mol for benchmark dimer binding energies) but low in cost (linear scaling with cluster size, if run in "embarrassingly parallel" mode). This approach employs flexible monomer units described at a correlated level of theory.

2. GENERALIZED MANY-BODY EXPANSION

Our interest in the MBE began as an attempt to understand the connections between the litany of fragment-based quantum chemistry methods developed over the past 15 years (12), and we have recently introduced an "elemental" classification scheme for such methods *(37)*, which helps to bring some semblance of order to this zoo of techniques. Our classification scheme is based upon a unifying, universal energy formula that arises from a GMBE that encompasses both the case where two or more fragments share nuclei in common and the case where all of the fragments are disjoint. In the latter case, the GMBE is equivalent to the traditional MBE (Equation 2.1). The rigorous derivation of the GMBE, outlined in Section 2.1, provides additional justification for existing methods based on overlapping fragments *(29,33,55)*, which previously had been introduced in a more *ad hoc* way, but also suggests new methods. In particular, numerical results in Section 2.2 demonstrate that two-body expansions based on overlapping fragments provide a route to very high accuracy fragment-based calculations in large systems.

2.1. Theory

At the heart of the GMBE is an expression for the ground-state energy for a system of N fragments,

$$E = E^{(1)} + \Delta E^{(2)} + \cdots + \Delta E^{(N)}, \tag{2.4}$$

written in the form of sequential n-body corrections to the one-body energy (sum of the fragment energies), $E^{(1)}$. Whatever its detailed form, we can at least state that the n-body correction, $\Delta E^{(n)}$, ought to consist of all interactions that occur in a system composed of n fragments but which are absent in a system composed of $n - 1$ fragments. Thus

$$\Delta E^{(n)} = E^{(n)} - E^{(n-1)}, \tag{2.5}$$

where $E^{(n)}$ represents the n-body approximation to the ground-state energy.

To understand what $E^{(n)}$ should look like, it helps to think of the fragments as sets of atoms, in a mathematical sense. Number the fragments with $I = 1, \ldots, N$ and denote the Ith fragment as $F_I^{(1)}$. One can then form dimers, trimers, $\ldots$, n-mers of fragments by taking the union of 2, 3, $\ldots$, n fragments. The number of n-mers is

$$\binom{N}{n} = \frac{N!}{n!(N-n)!} \tag{2.6}$$

and we will denote them as $F_J^{(n)}$, where $J = 1, 2, \ldots, \binom{N}{n}$. Let this (over-lapping) set of objects, $\{F_J^{(n)}\}$, constitute a new set of fragments. We have shown *(38)* that the Hamiltonian for the original (super)system can be expressed *exactly* in terms of the Hamiltonians of each n-mer along with the Hamiltonians of subsystems constructed from intersections of n-mers:

$$\hat{H} = \sum_{I=1}^{M} \hat{H}\left(F_I^{(n)}\right) - \sum_{I=1}^{\binom{M}{2}} \sum_{J<I} \hat{H}\left(F_I^{(n)} \cap F_J^{(n)}\right)$$

$$+ \sum_{I=1}^{\binom{M}{3}} \sum_{J<I} \sum_{K<J} \hat{H}\left(F_I^{(n)} \cap F_J^{(n)} \cap F_K^{(n)}\right) + \cdots. \tag{2.7}$$

Here, $M = \binom{N}{n}$ and $\hat{H}(S)$ denotes the Hamiltonian for set S. The expansion terminates naturally whenever the intersection $F_I^{(n)} \cap F_J^{(n)} \cap \cdots$ becomes empty. The expectation value of this Hamiltonian, computed using the supersystem's ground-state wave function (if it were known), would afford the exact ground-state energy. In other words, in the partition of the supersystem into overlapping fragments, we have not lost any of the one- or two-particle operators in the Hamiltonian, nor have we picked up any spurious ones, provided that we construct all of the non–empty intersections that are indicated in Equation (2.7).

The ground-state wave function, $|\Psi\rangle$, is not known in practice, so the approximation consists of replacing $|\Psi\rangle$ in $\langle\Psi|\hat{H}|\Psi\rangle$, where $\hat{H}$ has the structure given in Equation (2.7), with a direct product of localized wave functions for individual subsystems. In other words, we make approximations such as

$$\left\langle\Psi\left|\hat{H}\left(F_I^{(n)} \cap F_J^{(n)}\right)\right|\Psi\right\rangle \approx \left\langle\Psi_{I\cap J}^{(n)}\left|\hat{H}\left(F_I^{(n)} \cap F_J^{(n)}\right)\right|\Psi_{I\cap J}^{(n)}\right\rangle. \tag{2.8}$$

Here, $|\Psi_{I\cap J}^{(n)}\rangle$ is the ground-state wave function for the subsystem that is constructed from the intersection of n-mers $F_I^{(n)}$ and $F_J^{(n)}$. Essentially, Equation (2.8) is an appeal to Kohn's principle of the "nearsightedness of electronic matter" *(56,57)*, which in the end is the reason that fragment-based methods work at all.

Approximations as in Equation (2.8), in conjunction with the Hamiltonian in Equation (2.7), provide an expression for the n-body approximation $E^{(n)}$ to the ground-state energy of the supersystem. It is convenient to express this energy formula as

$$E^{(n)} = \sum_{I=1}^{\binom{N}{n}} \varepsilon_I^{(n)}, \tag{2.9}$$

in which $\varepsilon_I^{(n)}$ is the *intersection-corrected energy* of the n-mer $F_I^{(n)}$, defined as

$$\varepsilon_I^{(n)} = E_I^{(n)} - \sum_{J<I} E_{I\cap J}^{(n)} + \sum_{J<I}\sum_{K<J} E_{I\cap J\cap K}^{(n)} - \cdots \tag{2.10}$$

The quantity $E_{I\cap J}^{(n)}$ in this equation is the energy of a subsystem constructed from the nuclei common to $F_I^{(n)}$ and $F_J^{(n)}$. The signs alternate, and the expansion naturally terminates once there are no longer any non-empty intersections $F_I^{(n)} \cap F_J^{(n)} \cap \cdots$.

To digest all of this, let us consider what is involved in performing a GMBE-based calculation. First, one must somehow partition the nuclei into fragments ($F_I^{(1)}$), which may or may not overlap. Second, one selects an n-body level at which to truncate the GMBE and forms all n-mers of fragments. These n-mers are the quantities indexed by $I, J, \ldots$ in Equations (2.9) and (2.10). Finally, one must compute all of the mutual intersections amongst these n-mers, performing an electronic structure calculation on each n-mer (to obtain the $E_I^{(n)}$ terms in Equation 2.10) and also each non-empty intersection (to obtain the remaining terms).

This procedure lends itself to a systematic classification scheme for fragment-based quantum chemistry methods, in which four things must be specified *(37)*: first, the order n at which the GMBE will be truncated; second, a method for partitioning the supersystem's nuclei into subsystems (fragments); third, a capping method in the event that fragmentation severs any covalent bonds; and finally, a method for performing electrostatic embedding, if desired.

With these ideas in mind, we note that the GMBE has several appealing features. First, it recovers the exact ground-state energy when $n = N$. As with the traditional MBE, this result is not useful in practice, but does demonstrate that the various n-body approximations have a well-defined limit. In practice, it is reasonable to expect that the n-body corrections will get smaller as n increases, for essentially the same reasons that one expects this in the case of the traditional MBE. Finally, it can be shown that Equation (2.4), truncated at the n-body level, yields the same energy formula as n-body truncation of the traditional MBE, if the fragments are completely disjoint (containing no nuclei in common) *(37,38)*.

The GMBE is also closely connected to several other fragment-based methods that utilize overlapping fragments, although the GMBE is better justified and more complete, in our opinion. Alternative methods that exploit overlapping fragments include the *cardinality-guided molecular tailoring approach* (CG-MTA) *(29–32)*, and the *generalized energy-based fragmentation* (GEBF) method *(33–35)*. Only recently was it recognized that these two methods use precisely the same energy expression *(22)*, which amounts to a one-body truncation of the GMBE *(37)*. (This illustrates the important conceptual advantage of possessing a unifying energy formula.) As applied to non-covalent clusters, the *fragmentation* schemes employed in CG-MTA and GEBF are different, although both place multiple monomer units into a single fragment. As such, some many-body effects are incorporated despite the one-body truncation of the GMBE.

The energy formula obtained by truncation at $n = 1$ is also reminiscent of the energy formula used in the *systematic molecular fragmentation* (SMF) procedure developed by Collins and co-workers *(55,58–60)*. However, SMF generates only a limited subset of the intersections in Equation (2.10), and we emphasize that for set-theoretical reasons, all of these terms are required in order to avoid double-counting or omitting any of the one- and two-particle operators in the supersystem Hamiltonian. In contrast to our set-theoretical derivation of the GMBE *(37,38)*, the GEBF and SMF approaches invoke elaborate, *ad hoc* rules in an attempt to avoid serious overcounting. These rules ultimately lead to the coefficients of ± 1 that appear in Equation (2.10), although in the case of SMF, certain intersections are absent, and this roundabout way of obtaining the final result may account for the delayed recognition of the similarities between these methods.

2.2. Benchmark examples

Whereas previous quantum chemistry methods based on overlapping fragments amount to truncations of the GMBE at the one-body level, possibly with additional approximations, our more general formulation positions us to explore the effects of incorporating higher-order terms. To date, we have considered methods based on dimers of overlapping fragments *(37,38)* and have compared these to other GMBE-based methods. Initial applications have focused on exploring what is required in order to reproduce total energies to an accuracy of $\lesssim 1$ kcal/mol, with respect to a supersystem calculation performed using the same electronic structure method and basis set. A variety of non-covalent clusters have been explored *(37,38)*.

A quick survey of how various methods compare can be obtained from results for water clusters, and we will use these clusters to assess several different fragmentation schemes. The first of these uses a single H_2O molecule per fragment, corresponding to a traditional version of the MBE. We use the nomenclature MBE(2) to refer to the two-body version of this approach. (One-body truncation of the MBE is obviously not a useful method in the case of disjoint fragments.) The electrostatically embedded version of MBE(2) will be denoted EE-MBE(2), and following Dahlke and Truhlar (25) we use gas-phase Mulliken charges for H_2O to accomplish the embedding. (This simple embedding approach often proves to be remarkably robust (61), although one can imagine cases where it might fail and we have begun to explore more sophisticated, self-consistent charge-embedding schemes (37).)

We also consider a fragmentation scheme based on a distance criterion of $R = 3$ Å, as in some previous GEBF studies of water clusters (33–35). In this case, fragmentation proceeds by looping over oxygen nuclei. For oxygen nucleus I, we create a fragment $F_I^{(1)}$, consisting of all other oxygen nuclei within a distance R along with all covalently-bonded hydrogen nuclei. This procedure results in three to four H_2O molecules per fragment, meaning that it *does* make sense to consider a one-body truncation of the GMBE, since intermolecular interactions between nearby H_2O molecules are already included in the one-body energies. We consider both one- and two-body truncations of the GMBE for this fragmentation scheme, denoting these as GMBE(n) with $n = 1$ or 2, or EE-GMBE(n) when electrostatic embedding is employed. (The choice of gas-phase Mulliken embedding is probably not optimal in the case of overlapping fragments, where the polarization environment may be very different in different fragments, but is used here for simplicity. See Ref. (37) for a discussion of this point.) Note that GMBE(2) is distinct from MBE(2). In the latter method, no single calculation larger than $(H_2O)_2$ is required, whereas GMBE(2) with an $R = 3$ Å criterion for fragment formation may require calculations as large as $(H_2O)_8$.

Total errors in the electronic energies of small water clusters are shown in Figure 2.1 for each of the aforementioned methods. It is immediately evident from the data in Figure 2.1 that the MBE(2) approach is unacceptable (errors > 1 kcal/mol/monomer) unless electrostatic embedding is used, in which case the error is reduced to <0.5 kcal/mol/monomer. This is consistent with the importance of many-body polarization effects in polar systems and is essentially a recapitulation of previous work by Dahlke and Truhlar (25). The method that we call GMBE(1) is equivalent to what has been called GEBF in the previous literature (33), and with electrostatic

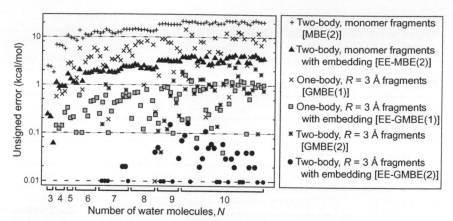

Figure 2.1 Total unsigned errors in GMBE calculations applied to small water clusters, relative to supersystem calculations using the same electronic structure model, B3LYP/ 6-31+G(d,2p). Consult the text for an explanation of the nomenclature. *Adapted from Ref. (37); copyright 2012 American Institute of Physics.*

embedding, this approach can reduce the *total* error to $\lesssim$ 1–2 kcal/mol in most cases. Errors per H_2O monomer are reduced to <0.1 kcal/mol/monomer. For the two-body method based on dimers of overlapping fragments, the error is <1 kcal/mol/monomer even without electrostatic embedding.

One might reasonably argue that the overlapping dimers used in these GMBE(2) calculations are not significantly smaller than the actual system size, and therefore, larger systems are needed to discover the intrinsic accuracy of these methods. For this purpose, we have used several configurations of a $(H_2O)_{57}$ cluster and several isomers of $F^-(H_2O)_{10}$. (Fluoride–water clusters, in particular, are a very challenging test case for fragment-based methods *(37,38,53,62)* and also for DFT *(54)*.)

Mean absolute errors (with respect to supersystem calculations) for the water clusters and fluoride–water clusters, using several different GMBE-based methods, are listed in Table 2.1. Unlike the data in Figure 2.1, the errors in Table 2.1 are computed on a per-monomer (i.e. per-H_2O molecule) basis, as one can argue that the error in fragment-based calculations for non-covalent systems is likely an extensive quantity *(19,51)*. The data clearly indicate the superiority of methods based on overlapping fragments, namely, the GMBE(n) and EE-GMBE(n) methods. With electrostatic embedding, even $n=1$ truncation of the GMBE (i.e. EE-GMBE(1)) affords sub-kcal/ mol accuracy for both $(H_2O)_{57}$ and the strongly-interacting $F^-(H_2O)_{10}$ system. Since the $R=3$ Å fragmentation criterion used here results in three to

Table 2.1 Mean absolute errors per H_2O monomer (in kcal/mol) in GMBE total energies, as compared to a supersystem calculation at the same level of theory [B3LYP/6-31+G(d,2p)]

Method	$(H_2O)_{N \leq 10}$	$(H_2O)_{57}$	$F^-(H_2O)_{10}$
MBE(2)	1.75	3.74	2.21
EE-MBE(2)	0.27	0.36	0.10
GMBE(1)	0.56	2.95	0.71
EE-GMBE(1)	0.06	0.40	0.74
GMBE(2)	0.06	0.02	0.08
EE-GMBE(2)	0.00	0.00	0.01

Data are taken from Refs. *(37,38)*. The MBE(2) methods are two-body expansions with monomer fragments, whereas the GMBE(*n*) methods use a distance criterion of $R = 3$ Å to construct fragments, leading to three to four H_2O monomers per fragment.

four H_2O molecules per fragment, the GMBE(1) method captures intermolecular interactions that do not appear until $n = 4$ in the traditional MBE.

Most significantly, we note that the GMBE(2) method, which uses overlapping dimers with three to four H_2O molecules per monomer, yields essentially exact results (as compared to a supersystem energy calculation), even without electrostatic embedding. Admittedly, GMBE(2) calculations require individual electronic structure calculations on subsystems containing as many as eight H_2O molecules, which is certainly not a huge calculation in itself but is much larger than any of the individual calculations required at the EE-GMBE(1) level, which are limited by the fragment size to at most four H_2O molecules in any given calculation. We are presently performing systematic tests to establish the accuracy *per unit cost* of various fragment-based approaches, but in terms of the accuracy it appears that GMBE(2) can effectively replace supersystem calculations.

The excellent performance of methods based on overlapping fragments is also significant as one contemplates extending the GMBE to macromolecular systems, where fragmentation must sever covalent bonds. Methods based on non-overlapping fragments are likely to be more susceptible to "edge effects" caused by the extra atoms ("double link atoms" *(63)*) introduced to cap severed valencies. Indeed, an alternative "many overlapping body" expansion (MOBE), introduced recently by Mayhall and Raghavachari *(36)*, shows great promise for macromolecular systems in preliminary tests. As compared to the GMBE, the MOBE introduces some—but not all—of the intersections that are required by the set-theoretical inclusion/exclusion principle upon which Equation (2.7) is based *(38)*. However, our own preliminary testing reveals that for strongly-interacting systems such as $F^-(H_2O)_6$ and $^+NH_3CH_2CO_2^-(H_2O)_{10}$ (zwitterionic glycine in water), the MOBE is more sensitive to the particulars of how the system is fragmented than is the GMBE *(38)*. The GMBE, truncated at the $n=2$ level, consistently affords small errors for these challenging test cases, even when the fragments are chosen in an unreasonable way, by selecting H_2O molecules at opposite ends of the cluster to be in the same fragment. This apparent robustness is an appealing feature of the GMBE as one imagines moving toward macromolecular systems, as is the fact that we have selected fragments based on a distance criterion that is to some extent systematically improvable. Testing of this approach for larger systems is currently underway in our group.

3. SYMMETRY-ADAPTED PERTURBATION THEORY FOR CLUSTERS

Whereas MBE- and GMBE-type approaches are essentially combinatorial in nature, we next describe a family of monomer-based methods, intended for computing non-covalent interactions in sizable systems, that are designed to exploit the physical nature of such interactions. These methods describe many-body polarization effects in a self-consistent yet linear-scaling way, using a form of the variational explicit polarization (XPol) method *(50)*, but then supplementing this with a pairwise-additive form of SAPT *(14–16)*. As with the GMBE-based methods discussed in the previous section, the latter step is embarrassingly parallelizable, since the individual SAPT calculations are completely independent of one another, and is thus amenable to large collections of monomers. Below, we introduce the combined XPol + SAPT methodology, which in principle establishes a systematically improvable hierarchy of *ab initio* methods for the description of non-covalent interactions, although in practice only the lowest rung of that hierarchy has been implemented. Nevertheless, this implementation extends SAPT methodology from dimers to many-body systems in an affordable way. Later, in Section 4, we discuss introducing some empiricism into this approach, which improves the results while greatly reducing the cost.

3.1. Theory

3.1.1 Monomer-based many-body polarization

The XPol method *(50)* is an approximate, fragment-based molecular orbital (MO) method that was originally developed with the intention that it would constitute a quantum chemistry-based polarization term for a "next-generation" force field *(64,65)*. In our hands, it serves as an efficient means by which many-body polarization (induction) effects can be folded into monomer-based SCF calculations. In a cluster, this will provide appropriate zeroth-order wave functions for a subsequent SAPT calculation, whereas a gas-phase starting point would omit crucial many-body polarization effects.

The XPol method is based on a direct-product *ansatz* for the supersystem wave function *(50,66)*,

$$|\Psi_{\mathrm{XPol}}\rangle = \prod_{A=1}^{N} |\Psi_A\rangle. \tag{2.11}$$

The individual monomer SCF wave functions, $|\Psi_A\rangle$, are each properly antisymmetric with respect to exchange of electrons within monomer A, but intermolecular exchange effects are neglected at this level. In the original XPol method of Xie *et al.* *(50,64,65)*, exchange-repulsion and dispersion interactions were modelled using empirical intermolecular potentials of Lennard–Jones or Buckingham type. This approach, however, proves to be inadequate for high-accuracy binding energy calculations; we find *(54)* a mean unsigned error of 3.2 kcal/mol for the hydrogen-bonded subset of the S22 database *(67,68)*, despite the fact that the Lennard–Jones parameters were optimized for H-bonded systems *(69)*. In the XPol+SAPT approach, we sidestep the need for these empirical potentials by means of SAPT, as described below. In effect, we utilize XPol to generate monomer SCF wave functions, $|\Psi_A\rangle$, whose MOs implicitly contain many-body polarization effects and which are appropriate starting points for many-body SAPT.

Omitting empirical potentials, and assuming closed-shell monomers for simplicity, the XPol energy can be written as *(50,51)*

$$E_{\mathrm{XPol}} = \sum_{A=1}^{N} \left[2\sum_{a\in A} \mathbf{c}_a^\dagger \left(\mathbf{h}^A + \mathbf{J}^A - \frac{1}{2}\mathbf{K}^A \right) \mathbf{c}_a + E_{\mathrm{nuc}}^A \right] + E_{\mathrm{embed}}. \tag{2.12}$$

The term in square brackets is the ordinary HF energy expression for monomer A, but we assume that the MOs $\mathbf{c}_a$ for monomer A are expanded using atom-centred Gaussian basis functions located on monomer A only, in what has been termed the *absolutely localized* MO (ALMO) basis *(70)*. This affords a method whose cost grows only linearly with respect to the number of monomers, N, and furthermore excludes basis set superposition error (BSSE) by construction. In compact basis sets, the ALMO *ansatz* also excludes charge transfer between the monomers *(70)*.

The final term in Equation (2.12) is an electrostatic embedding potential, E_{embed}, by means of which the monomers $B \neq A$ make their presence felt during the SCF iterations for monomer A. One can envision various ways to accomplish this embedding; the "correct" way (corresponding to exact Coulomb interactions within the ALMO *ansatz*) is to use the electrostatic potential arising from the SCF electron densities ρ_B. However, this requires two-electron Coulomb integrals between two monomers, and in the interest of reducing the computational cost, XPol calculations instead collapse ρ_B onto atom-centred point charges, so that only one-electron charge-density integrals are required. In the original implementation of XPol by Xie *et al.* *(50,64,65)*, Mulliken charges were used for this purpose. This choice is fine in the context of the minimal-basis, semi-empirical electronic structure methods used in those studies, but is problematic in larger basis sets, often leading to SCF convergence failure *(51)*. As an alternative, we use CHELPG charges *(71)*, which are derived from a least-squares fit of ρ_B's electrostatic potential, evaluated on a grid *(51,52)*. (Since XPol is ultimately designed as a simulation method, we have developed a special smooth implementation of the CHELPG charges for this purpose; see Ref. *(52)*.) Note that this aspect of the methodology is systematically improvable; higher-order terms in the multipole expansion could be incorporated, although in practice the dipole moment arising from the CHELPG charges is often in reasonable agreement with the SCF dipole moment.

Variational minimization of Equation (2.12) with respect to the $\mathbf{c}_a \in A$, subject to the constraint that the MOs within monomer A remain orthonormal, affords an XPol SCF equation for monomer A *(50)*:

$$\mathbf{F}_A \mathbf{C}_A = \mathbf{S}_A \mathbf{C}_A \varepsilon_A. \tag{2.13}$$

The Fock matrix for monomer A can be expressed in the atomic orbital basis as *(51)*

$$(\mathbf{F}_A)_{\mu\nu} = (\mathbf{f}_A)_{\mu\nu} - \frac{1}{2} \sum_{J \notin A} \left(\mu \left| \frac{q_J}{|\vec{r} - \vec{R}_J|} \right| \nu \right)_{\vec{r}} + \sum_{J \in A} \left(\frac{\partial E_{\text{embed}}}{\partial q_J} \right) \left(\frac{\partial q_J}{\partial (\mathbf{P}_A)_{\mu\nu}} \right). \tag{2.14}$$

The first term on the right is the gas-phase Fock matrix, the second term represents Coulomb interactions between point charges q_J (centred on monomers $B \neq A$) with the SCF density ρ_A, while the third term captures how A's own point charges vary as a function of its density matrix, $\mathbf{P}_A$.

Because the embedding charges for monomers $B \neq A$ show up in the Fock matrix for A, a "dual SCF" procedure is required *(50,51,64)*, with an outer loop over $A = 1, 2, \ldots, N$ and an inner loop that iterates Equation (2.13) to self-consistency for monomer A. Note that the final term in the XPol Fock matrix ensures that the converged XPol energy is fully variational *(50)*; explicit expressions for this term can be found in Ref. *(51)* for the case of Mulliken, Löwdin, or CHELPG embedding charges.

3.1.2 Symmetry-adapted perturbation theory

Our original idea for the XPol + SAPT methodology went something like this *(51)*: motivated by the large many-body polarization effects observed in clusters of polar molecules, but the (nearly) pairwise-additive nature of other intermolecular interactions, we should use XPol to obtain monomer wave functions that are better approximations for the cluster environment than would be gas-phase wave functions, and then use these as zeroth-order states for a pairwise-additive SAPT calculation. Subsequently, however, we showed how to arrive at almost the same methodology via systematic approximations to the Schrödinger equation for the cluster *(52)*, and we will outline the latter derivation here. First, we write the Hamiltonian for the cluster in terms of gas-phase Fock operators $(\hat{f}_A)$ and Møller–Plesset fluctuation operators $(\hat{W}_A)$ for monomer A, along with intermolecular interaction operators $(\hat{V}_{AB})$:

$$\hat{H} = \sum_A \hat{f}_A + \sum_A \xi_A \hat{W}_A + \sum_A \sum_{B<A} \zeta_{AB} \hat{V}_{AB}. \qquad (2.15)$$

The quantities ξ_A and ζ_{AB} are parameters to count orders in perturbation theory. The zeroth-order wave function, $|\Psi_0\rangle$, is the direct product in Equation (2.11), which is consistent with a zeroth-order Hamiltonian of the form

$$\hat{H}_0 = \sum_{A=1}^{N} \hat{f}_A. \qquad (2.16)$$

The zeroth-order energy, E_0, is then the sum of all occupied MO eigenvalues.

The partitioning in Equation (2.15) leads to the following expression for the interaction energy $E_{\text{int}} = E - E_0$:

$$E_{\text{int}} = \sum_A \xi_A \langle \Psi_0 | \hat{W}_A | \Psi \rangle + \sum_A \sum_{B<A} \zeta_{AB} \langle \Psi_0 | \hat{V}_{AB} | \Psi \rangle, \qquad (2.17)$$

where $| \Psi \rangle$ is the exact wave function for the entire system. In order to construct an intermolecular perturbation theory, one can expand the interaction energy and wave function in terms of the N parameters ξ_A and $N(N-1)/2$ parameters ζ_{AB}. In our work *(51,52)*, we have investigated only a small number of these terms. This includes terms that are first-order in the intramolecular corrections, for which the corresponding energy corrections are simply expectation values of the fluctuation operators $\hat{W}_A$ with respect to zeroth-order wave functions.

The more interesting terms are intermolecular in nature; we consider only the intermolecular terms that neglect monomer electron correlation (all $\xi_A = 0$), through second order in perturbation theory ($\sum_A \sum_{B<A} \zeta_{AB} \leq 2$). In the context of traditional SAPT calculations, this approximation is often termed SAPT0 *(15)*, but we will call it SAPT(HF) in order to maintain consistency with the notation SAPT(KS). Both methods are based on the same perturbation formulas, but SAPT(KS) uses MOs and eigenvalues obtained from Kohn–Sham (KS) DFT. Three types of corrections satisfy these criteria: first-order terms where a single $\zeta_{AB} = 1$ (with all others zero), second-order terms with $\zeta_{AB} = 2$, and second-order terms where $\zeta_{AB} = \zeta_{BC} = 1$. The first two cases result in pairwise-additive electrostatic, induction, and dispersion corrections, whereas the third cases couples the induction amplitudes on two different pairs of monomers. We refer to the latter case as *three-body induction couplings (52)*.

Equation (2.17) looks like standard Rayleigh–Schrödinger perturbation theory (RSPT), albeit with a large number of separate perturbations. The RSPT approach neglects the antisymmetry requirement of the wave function, which is important at short range where the monomer wave functions overlap. Following Jeziorski et al. *(14)*, we instead apply *symmetrized* RSPT, in which we accept non-antisymmetric corrections to the direct-product wave function but then project away any Pauli-forbidden wave function components in the energy corrections. This amounts to a modification of the interaction energy formula in Equation (2.17). The symmetry-adapted formula is

$$E_{\text{int}} = \left[\sum_A \xi_A \langle \Psi_0 | \hat{W}_A \hat{A} | \Psi \rangle + \sum_A \sum_{B<A} \zeta_{AB} \langle \Psi_0 | \hat{V}_{AB} \hat{A} | \Psi \rangle \right] \langle \Psi_0 | \hat{A} | \Psi \rangle^{-1},$$

$$(2.18)$$

where $\hat{A}$ is an antisymmetrizer. Order-by-order energy corrections can be obtained in a manner analogous to the RSPT expressions discussed above, and each term in the RSPT expansion is found to have a corresponding exchange term.

The exchange corrections derived from Equation (2.18) are quite complex and we have simplified them by utilizing the *single-exchange approximation (14,72,73)*, wherein all permutations appearing in $\hat{A}$ that involve exchange of more than one pair of electrons are neglected. In addition, we have introduced a "diagonal-exchange" approximation (see Ref. *(52)* for details). Expressions for higher-order exchanges have been derived in the context of traditional SAPT *(72–74)*, and there is no reason in principle why these cannot be included in the XPol + SAPT methodology, albeit at increased cost and complexity.

3.1.3 Combining XPol with SAPT

We start with some nomenclature. In the previous work *(51–53)*, we used the acronym "XPS" to refer to the combined XPol + SAPT methodology, but this runs the risk of being confused with X-ray photoelectron spectroscopy, also commonly called XPS. As such, we henceforth use the acronym "XSAPT" as an abbreviation for XPol + SAPT and will use the notation XSAPT (*method*) to refer to XSAPT calculations where *method* (which might be HF theory or some KS density functional) is used to generate monomer wave functions and orbital eigenvalues.

Returning to the theory, note that the intermolecular perturbation indicated in Equations (2.17) and (2.18) is valid only for non-interacting zeroth-order monomers, that is, for traditional SAPT calculations. When XPol is used to generate the zeroth-order wave functions, the perturbation used in SAPT must be slightly modified to avoid double-counting intermolecular interactions *(51)*. Following this modification, the total XSAPT energy is defined to be the sum of zeroth-order fragment energies (HF eigenvalues), first-order intramolecular RSPT corrections of the form $\langle \Psi_0 | \hat{W}_A | \Psi_0 \rangle$, and both first- and second-order RSPT and SAPT corrections. The resulting energy expression can be written as *(52)*

$$
\begin{aligned}
E_{XSAPT} = \sum_A & \left(\sum_{a \in A} \left[2\varepsilon_a^A - \mathbf{c}_a^\dagger \left(\mathbf{J}^A - \frac{1}{2}\mathbf{K}^A \right) \mathbf{c}_a \right] + E_{nuc}^A \right) \\
& + \sum_A \sum_{B<A} \left[E_{RSPT}^{[0;1_{AB}]} + E_{exch}^{[0;1_{AB}]} + E_{RSPT}^{[0;2_{AB}]} + E_{exch}^{[0;2_{AB}]} \right] \\
& + {\sum_C}' {\sum_{D<C}}' \left(E_{RSPT}^{[0;1_{AB},1_{CD}]} + E_{exch}^{[0;1_{AB},1_{CD}]} \right),
\end{aligned}
\tag{2.19}
$$

where the superscripts denote the various energy corrections. (Due to a typographical error, the exchange terms in the second line are absent from the corresponding expression in Ref. *(52)*.) The primes on the final set of summations indicate that these include only terms with at least one index in common between AB and CD, as a result of the diagonal-exchange approximation. Note that the term involving summation over a is simply the HF energy expression for monomer A in the absence of charge embedding.

Without any formal justification, we have extended the XSAPT methodology to a KS–DFT description of the monomers simply by adding the DFT exchange-correlation contribution to the gas-phase Fock matrix $\mathbf{f}_A$ for each monomer. In the context of traditional SAPT calculations, this extension is known as SAPT(KS) *(75)*. Although this represents a tempting way to incorporate intramolecular electron correlation at low cost, the perils of SAPT(KS)—namely, severe overestimation of dispersion interactions due to incorrect asymptotic behaviour of the exchange-correlation potential—are well known *(76)*. These problems are inherited by our XSAPT(KS) generalization to many-body systems, as discussed below.

Regarding notation, we refer to the above methodology as either XSAPT(HF) or XSAPT(KS) depending on how the monomer MOs were generated. In addition, we will show some results computed using XSAPT (KS)-resp, where "resp" denotes a response correction *(51)*. In this approach, the second-order induction and exchange-induction amplitudes are replaced by amplitudes obtained by solving coupled-perturbed Hartree–Fock (CPHF) equations *(77)* for each monomer in each $A\cdots B$ dimer, where the perturbation is the electrostatic potential of the other monomer. This affords an infinite-order correction for polarization arising from a frozen partner density *(78)*.

Finally, some discussion of basis sets is warranted. Typically, SAPT calculations are performed in the *dimer-centred basis set*, meaning that the combined $A+B$ basis set is used to calculate the zeroth-order wave functions for both A and B. This tends to give superior results as compared to using only a monomer basis set *(79)*, but is incompatible with the ALMO *ansatz* that is used in XPol. Instead, we converge monomer wave functions using monomer basis sets, but perform the dimer SAPT calculations in a what we have termed a "pseudocanonicalized" or "projected" basis set *(51)*. In this approach, we first converge XPol SCF equations in the monomer basis sets, for efficiency, and then construct XPol Fock matrices for A and B in the dimer ($A+B$) basis set. The occupied–occupied and virtual–virtual blocks of these matrices are then separately diagonalized, which leaves the fragment densities and zeroth-order fragment energies unchanged. (This idea, which has been called *pseudocanonicalization*, is also used in dual-basis MP2 calculations *(80,81)*.) As compared to the monomer basis set, this approach provides a larger set of virtual orbitals that extend over the partner fragment (and can thus capture intermolecular charge transfer), and we use this larger virtual space to evaluate the perturbative corrections. Note, however, that the pseudocanonical MOs are no longer eigenfunctions of the fragment Fock matrices, since the occupied–virtual blocks of those matrices are non-zero. In principle, we could include a perturbative correction to the zeroth-order energies, of the form $\sum_{ar}(\mathbf{F}_A)_{ar}/(\varepsilon_a - \varepsilon_r)$ for fragment A. We decline to do so, however, as this would introduce BSSE.

3.2. Numerical tests

Unlike GMBE-based methods, XSAPT is not intended to reproduce or approximate the results of any particular theoretical model chemistry. Instead, our goal is to construct a systematically improvable methodology for non-covalent interactions—in clusters, molecular liquids, and solids— that is affordable enough to be used in *ab initio* molecular dynamics and/ or Monte Carlo simulations. With this in mind, we adopt a pragmatic approach that uses basis sets of moderate size only, and low-order perturbation theory, but always benchmarking against the best available theory. We shall see that the choice of basis set is crucial, as there is a delicate basis-set dependence to how well XSAPT describes electrostatic and induction interactions versus dispersion interactions.

The first high-level benchmarks to which we compare are dimer binding energies from the S22A database *(68)*, which is a slight revision to the complete basis-set (CBS) CCSD(T) energetics of the original S22 database assembled by Hobza and co-workers *(67)*. Many-body effects are of course absent in these systems, and probably for that reason, the results are fairly insensitive to the charge-embedding scheme that is used in XPol; in fact, the choice of no charge embedding at all (i.e. traditional SAPT) yields similar performance to XSAPT for S22A binding energies *(51)*. At the very least, this demonstrates that the XPol procedure does not degrade the quality of SAPT, and furthermore, we gain some important information regarding the basis-set dependence of the methodology, as discussed below.

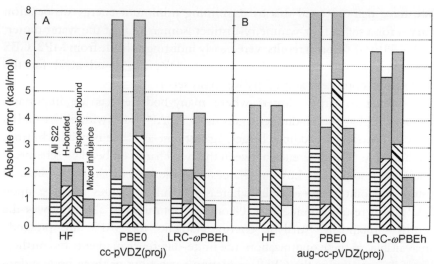

Figure 2.2 Mean absolute errors in SAPT(HF)-resp and SAPT(KS)-resp calculations for the S22A database. Results for two different density functionals and two different projected basis sets are shown. In each case, we plot the mean absolute error (open and hatched bars) and maximum absolute errors (grey bars) evaluated over the entire S22 set as well as three different subsets. For the SAPT(PBE0)/aug-cc-pVDZ(proj) calculations, the maximum error in the dispersion-bound subset is 11.5 kcal/mol, off-scale in the figure. *Adapted from Ref. (52) with permission from the PCCP Owner Societies.*

Figure 2.2 shows S22 error statistics for the SAPT(HF) method and also for SAPT(KS) with two different density functionals. For the dispersion-bound subset of S22, the SAPT(KS) errors are much larger than SAPT(HF) errors, which is known to result from the incorrect asymptotic behaviour of popular density functionals *(76)*. Range-separated functionals such as LRC-ωPBEh *(82)* improve—but do not completely eliminate—this behaviour, which results primarily from too-small energy denominators in the second-order, sum-over-states dispersion formula that is used in SAPT0. We note also that the addition of diffuse basis functions exacerbates this problem, as does enlarging the basis set in general *(52)*. We shall return to this point in Section 4.

Due to error cancellation, compact double-ζ basis sets, such as cc–pVDZ, afford reasonable errors of ~1 kcal/mol across the S22A data set *(51)*. Unfortunately, only augmented basis sets, combined with the aforementioned pseudocanonicalization technique, are able to provide accurate energetics for H-bonded systems across the full range of intermolecular distances.

We have demonstrated this by examining minimum-energy dissociation curves for a set of four symmetry-distinct isomers *(83)* of the water dimer; SAPT(HF)/aDZ(proj) results were nearly indistinguishable from MP2/CBS results *(51)*. (We use the abbreviations aDZ, aTZ, and aQZ for the correlation-consistent aug-cc-pVXZ basis sets.)

Looking to larger clusters where many-body effects are important, Figure 2.3 plots the binding energies using XSAPT(HF) with either CHE-LPG or Löwdin embedding charges, against best-available benchmarks from the literature. The difference between the two charge-embedding schemes is dramatic; results using Löwdin charges are little better than those obtained using no point charges at all, whereas CHELPG charges improve upon SAPT(HF) results considerably. Geometry optimization also improves the results substantially. Analytic energy gradients for SAPT are presently unavailable, so these optimizations use finite-difference gradients and further-more do not employ the CPHF induction corrections or three-body induc-tion couplings. In applications to $(H_2O)_6$, where finite-difference optimizations using these corrections are feasible, these additional corrections

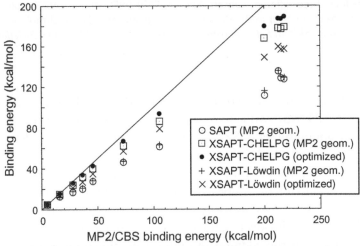

Figure 2.3 SAPT(HF)/aDZ(proj) and XSAPT(HF)/aDZ(proj) binding energies for $(H_2O)_n$ clusters, as compared to MP2/CBS benchmarks. For the XSAPT calculations, both the three-body induction couplings and the infinite-order response correction are neglected; results are shown both at the benchmark MP2 geometries and at geometries that have been self-consistently optimized (or at least relaxed; see Ref. *(51)*). The cluster isomers include four different $n = 20$ isomers along with one isomer each for $n = 2, 3, 4, 5, 6, 8,$ and 11. The diagonal line indicates where the (X)SAPT binding energy would coincide with the benchmark. *Adapted from Ref. (51); copyright 2011 American Institute of Physics.*

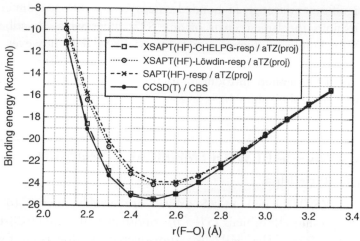

Figure 2.4 One-dimensional potential energy scans for $F^-(H_2O)$ along the $F \cdots O$ distance coordinate, at a fixed H_2O geometry. *Adapted from Ref. (52) by permission of the PCCP Owner Societies.*

are found to yield quantitative improvements in the energetics and, in at least one instance, a qualitative difference in the optimized geometries *(52)*.

In Ref. *(52)*, we investigated $X^-(H_2O)_n$ clusters (X=Cl, F, and OH) whose binding energies are much larger than those of any of the S22 dimers. The results provide useful information about basis-set dependence for strongly H-bonded clusters. Figure 2.4 shows that the progression

$$SAPT(HF) \rightarrow XSAPT(HF)\text{-Löwdin} \rightarrow XSAPT(HF)\text{-CHELPG}$$

yields substantial, progressive improvement in the binding energy of the challenging $F^-(H_2O)$ system. This indicates that the XPol charge embedding is able to capture self-consistent polarization in a reasonably accurate way, provided that the charges accurately represent the electrostatic potential, as CHELPG charges are designed to do.

We should point out that obtaining accurate results for $X^-(H_2O)_n$ requires augmented triple-ζ-quality basis sets. In general, for H-bonded systems described at the XSAPT(HF) level, the accuracy of binding energies increases systematically with basis-set quality *(52)*. Unfortunately, basis sets such as aTZ lead to substantial overestimation of dispersion interactions. Although we have suggested *(52)* def2-TZVP as a compromise choice that affords reasonable results in both H-bonded and dispersion-bound clusters, at low order in SAPT, these two types of systems are pushing the basis set demands in different directions. In addition, monomer correlation effects can be significant in strongly H-bonded systems (e.g. formic acid dimer *(84)*, which exhibits the largest error of any S22 dimer when described by SAPT0). Including monomer correlation via KS-DFT, however, only exacerbates the basis-set issues with the second-order SAPT0 treatment of dispersion. The method introduced in the next section is intended to circumvent these difficulties.

4. XPOL + SAPT WITH EMPIRICAL DISPERSION

4.1. Basic idea

Results of XSAPT based on KS orbitals are notably inferior to those obtained by using HF orbitals *(51–53)*, consistent with SAPT(KS) results in general *(75,76)*. The MP2-like sum-over-states dispersion formula already tends to overestimate dispersion energies at the SAPT0 level, and this tendency is exacerbated by KS orbitals, for which HOMO/LUMO gaps are generally smaller *(75)*. Using functionals with correct asymptotic behaviour, such as long-range corrected (LRC) functionals *(85–88)*, leads to wider gaps and reduces this tendency somewhat, as can be seen by comparing the PBE0 and LRC-ωPBEh results in Figure 2.2, but the results are still worse, in general, than those obtained using HF MOs.

    Ironically, the problematic dispersion and exchange-dispersion terms in SAPT(KS) are not only the least accurate ones but also the most expensive to compute, scaling as $\mathcal{O}(n^4)$ and $\mathcal{O}(n^5)$, respectively, with respect to monomer size, n. As such, we have recently introduced a method called XSAPT(KS) + D *(53)*, in which these problematic and expensive terms are replaced with empirical atom–atom dispersion potentials. Actually, this was first done in the context of SAPT(KS) by Misquitta and Szalewicz *(108)*, and later by Hesselmann *(89)*. We borrow Hesselmann's empirical potentials for this purpose:

$$E_{\text{disp}}^{\text{SAPT+D}} = -s_\beta \sum_{i \in A} \sum_{\substack{j \in B \\ (B \neq A)}} f_{\text{damp}}\left(r_{ij}\right) \frac{C_{ij}}{r_{ij}^\beta}. \tag{2.20}$$

Here, i and j represent nuclei located on different monomers, and

$$f_{\text{damp}}\left(r_{ij}\right) = \text{erf}\left(\frac{\alpha r_{ij}}{R_i + R_j}\right) \tag{2.21}$$

damps the empirical potential to zero at short distance. The parameters C_{ij} are defined in terms of atomic C_6 coefficients. These coefficients, along with the atomic radii R_i, are taken from the work of Grimme *(90)*. The parameters α, β, and s_β were fit to reproduce dimer binding energies *(53,89)*.

Similar empirical dispersion corrections have become quite popular in DFT *(9)*. As compared to such "DFT+D" methods, however, the empirical dispersion correction is better justified, formally speaking, in the SAPT+D case, since the dispersion terms are well defined and can be isolated in the perturbation expansion and replaced with empirical potentials. In contrast, parameterization of +D corrections for DFT+D must not only correct for DFT's faulty description of dispersion interactions but also subtract out whatever part of the dispersion interaction is already included in the functional, insofar as it was parameterized using experimental data *(53,89)*. Perhaps as a result, it is found to be necessary to use a rather different damping function in DFT+D, which decays much more slowly than that used in SAPT+D *(89)*.

In order to obtain a more accurate description of the monomers, we would like to use SAPT(KS) rather than SAPT(HF). As discussed in Section 3 and in the SAPT(KS) literature *(75,76)*, the use of asymptotically correct exchange-correlation potentials is mandatory in order to obtain reasonable results. However, the standard asymptotic correction schemes used in SAPT(KS) and SAPT(DFT) calculations *(91,92)* are ones in which correct asymptotic behaviour is grafted onto v_{xc} itself. Such approaches suffer from an inconsistency in that the corrected potentials v_{xc}^{AC} cannot be obtained as functional derivatives of any energy functional $E_{xc}[\rho]$. As an alternative, we use the LRC-ωPBEh functional *(82)* with a system-specific choice of the range-separation parameter, ω. As suggested by Baer *et al. (88)*, we "tune" ω so as to satisfy the condition

$$\varepsilon_{HOMO} = -IP \tag{2.22}$$

for the lowest ionization potential, computed using the same functional. (We also optimize the fraction of short-range HF exchange in LRC-ωPBEh, using S22A benchmarks *(53)*.)

The resulting method, which we call XSAPT(KS)+D, offers significant advantages over alternative electronic structure methods for non-covalent interactions *(53)*:

- Unlike XSAPT(HF), it incorporates intramolecular correlation, and in a relatively low-cost way.
- The expensive and inaccurate sum-over states dispersion formulas are replaced by simple scalar potentials.

- Unlike SAPT+D, the method is applicable to any number of monomers.
- The cost of XSAPT(KS)+D calculations scales as $\mathcal{O}(n^3)$ with respect to monomer size and scales as either $\mathcal{O}(N^2)$ or $\mathcal{O}(N^3)$ with respect to the number of monomers, depending upon whether three-body induction couplings are included or not. As such, this method is amenable to large systems and in particular is significantly faster than DFT+D.

4.2. Performance

We assess the accuracy of XSAPT(KS)+D against CCSD(T)/CBS benchmarks for dimer binding energies. Since the +D potentials contain parameters fitted against the S22 database, we will use the S66 database *(93)* for testing. (This database is also larger and thought to be more balanced in terms of various types of intermolecular interactions.) Figure 2.5 shows error statistics for a variety of methods applied to S66, including several different MP2- and CCSD-based methods extrapolated to the CBS limit. Notably, this list includes two "SCS(MI)" methods, which are spin-component-scaled (SCS) versions of MP2 and CCSD with scaling parameters for the same- and

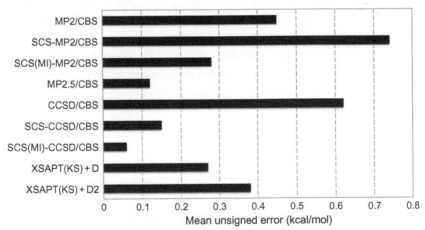

Figure 2.5 Error statistics with respect to CCSD(T)/CBS benchmarks for the S66 data set. The partially augmented aDZ′ basis set *(84,94)* (which is missing the diffuse functions on hydrogen and the diffuse *d* functions on the heavy atoms, as compared to aDZ) is used for XSAPT(KS)+D, and the "heavy augmented" ha-TZVPP basis set *(54)* is used for XSAPT(KS)+D2; other methods have been extrapolated to the CBS limit. In addition, the δE_{int}^{HF} correction *(15)* is used in the XSAPT(KS)+D2 calculations. *Data were obtained from Refs.* (53,54,93).

opposite-spin correlation energies that are specifically fit for molecular interactions, meaning that they are fit to S22 binding energies *(95,96)*.

XSAPT(KS)+D/aDZ′, where aDZ′ denotes a partially augmented version of the cc-pVDZ basis set *(84,94)*, outperforms all of the MP2-based methods listed in Figure 2.5, with a mean unsigned error (MUE) of just 0.27 kcal/mol and a maximum error <1 kcal/mol. Note that the MP2.5 method is a mixture of MP2 and MP3 *(97)*, hence all of the methods in Figure 2.5 that outperform XSAPT(KS)+D and +D2 are methods that exhibit sixth-order scaling with respect to supersystem size.

Results from the effective fragment potential (EFP) method *(98,99)* are also worthy of mention. The EFP is a classical, polarizable potential designed and parametrized specifically to reproduce *ab initio* results for intermolecular interactions *(99)*. When EFP-optimized geometries are used, the MUE for S66 binding energies predicted by the EFP method can be as low as 0.6 kcal/mol *(100)*. On the other hand, the error statistics shown in Figure 2.5 for *ab initio* and dispersion-corrected XSAPT methods are based on benchmark geometries computed at the RIMP2/cc-pVTZ level *(93)*. For the S22A data set, the MUE for EFP binding energies is 0.91 kcal/mol when evaluated using self-consistent EFP geometries, but rises to 1.79 kcal/mol when benchmark RIMP2/cc-pVTZ geometries are used instead *(100)*. In fact, the EFP method was designed only for rigid monomers *(98)*, and thus while EFP may be useful for rigid-monomer simulations (propagating dynamics on the EFP surface), it fails badly in cases where the actual monomer geometries are different from the ones used to parameterize the EFP.

On this note, it is important to recognize that for simulations, one wants an accurate potential energy *surface*, not just accurate binding energies. We find that XSAPT(KS)+D yields accurate potential energy curves for a variety of challenging systems, including Ar$\cdots$Ne, benzene dimer, formic acid dimer, and $Cl^-(H_2O)$ *(53)*. We do observe qualitative problems in $F^-(H_2O)$ at distances less than the sum of the atomic van der Waals radii *(53)*. At the time, we speculated that this might be caused by a breakdown of the single-exchange approximation at short range (a phenomena that we have observed in traditional SAPT calculations for this system *(62)*), or possibly due to the presence of a low-energy diabatic state corresponding to $FH\cdots OH^-$ *(101)*.

Very recently, however, we introduced a "second-generation" (+D2) version of this method that corrects this problem *(54)*. In this XSAPT(KS)+D2 method, we replace the empirical dispersion potential in Equation (2.20), which was parameterized to reproduce S22 binding energies using SAPT+D, with an alternative potential *(102)* that is parameterized to reproduce SAPT dispersion energies. In this sense, the +D2 correction is a "pure" dispersion potential.

The performance of XSAPT(KS)+D2 using the partially augmented ha-TZVPP basis set *(54)* is slightly worse for S66 as compared to XSAPT(KS)+D/aDZ′ results, but in fact, the second-generation method better reproduces the individual energy components (electrostatics, exchange, induction, and dispersion), as judged by comparison to high-level SAPT2+(3)/aTZ results *(54)*. This has allowed us to develop a many-body energy decomposition scheme based on XSAPT(KS)+D2 *(54)*. The results for the +D2 method suggest that the first-generation XSAPT(KS)+D method benefits from substantial cancellation of errors, but that XSAPT(KS)+D2 is more robust in this respect. Additional evidence in favour of this robustness is the fact that we can obtain such favourable error statistics using a triple-ζ basis set, whereas for the first-generation method, the results in triple-ζ basis sets are not nearly as good as they are for aDZ′ *(53)*. We find that triple-ζ basis sets are required for good performance in strongly H-bonded systems such as $F^-(H_2O)$.

We have also tested XSAPT(KS)+D on the same set of $(H_2O)_n$ clusters (up to $n=20$) that were considered in Figure 2.3. Figure 2.6 plots the binding energies obtained from various SAPT-based methods against MP2/CBS benchmarks. The XSAPT(KS)+D/aDZ′ results are significantly more accurate than pairwise-additive SAPT alone (which lacks the XPol

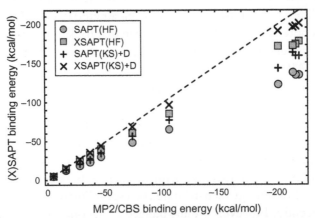

Figure 2.6 Binding energies for some $(H_2O)_n$ clusters, $n=2$–20, computed at MP2 geometries. Binding energies on the vertical axis were computed with various (X)SAPT methods using the aDZ′ basis, whereas the horizontal axis represents MP2/CBS benchmarks. Along the dashed line, the (X)SAPT result would be equal to the benchmark value. *Adapted from Ref. (53); copyright 2012 American Chemical Society.*

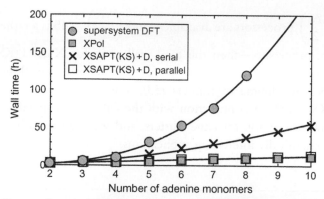

Figure 2.7 Timings for calculations on π-stacked (adenine)$_N$ oligomers. All calculations used the LRC-ωPBEh functional and the aDZ′ basis set. When run independently on $\binom{N}{2}$ processors, the requisite $\binom{N}{2}$ separate SAPT calculations represent a trivial amount of overhead on top of the (serial) XPol time, but even in serial mode (where all dimer SAPT calculations are run on the same processor), the overhead scales as $\mathcal{O}(N^2)$ with a small prefactor. *Adapted from Ref. (53); copyright 2012 American Chemical Society.*

treatment of many-body polarization), and also more accurate than XSAPT(HF), which lacks monomer electron correlation and also uses the sum-over-states dispersion formula.

Finally, to emphasize the significant computational advantage of XSAPT(KS)+D as compared, for example, to supersystem DFT+D, we present timings for calculations performed on (adenine)$_N$ systems up to $N=10$, in Figure 2.7 *(53)*. Three-body induction couplings are not included in the XSAPT(KS)+D results, which therefore scale as $\mathcal{O}(N^2)$, and the total wall time required for these calculations, even running in serial mode, is significantly less than what would be required for (less accurate) DFT+D calculations. The XPol cost scales as $\mathcal{O}(N)$, whereas supersystem DFT+D is $\mathcal{O}(N^3)$, and the actual XPol time for $N=10$ adenine units is only about four times larger than the wall time required for $N=2$. (This implies that the XPol scaling is actually slightly sublinear. Although this pleasing result is really just an artefact of the number of SCF cycles required in each case, it does at least suggest that the number of cycles required does not grow substantially as a function of supersystem size.) The subsequent $\mathcal{O}(N^2)$ step is embarrassingly parallelizable, and with the $\mathcal{O}(n^4)$ and $\mathcal{O}(n^5)$ steps eliminated from the SAPT calculations, the time required for any one of the dimer SAPT calculations is a small amount of overhead on top of the total XPol time. If $\begin{pmatrix} N \\ 2 \end{pmatrix}$ processors are available—and $\begin{pmatrix} 10 \\ 2 \end{pmatrix} = 45$ is a rather modest number of processors—then the total wall time for XSAPT(KS)+D in (adenine)$_N$ is nearly the same as the XPol time and scales linearly with the number of adenine monomers *(53)*. The sub-kcal/mol accuracy of XSAPT(KS)+D2, in conjunction with the calculation speed, makes this a very attractive option for large clusters, and we are presently pursuing a variety of chemically interesting applications.

5. CONCLUSIONS AND OUTLOOK

A purely *ab initio* description of intermolecular interactions is a challenging problem. Achieving an accuracy of $\lesssim 0.5$ kcal/mol typically requires either methods with triple excitations or else the introduction of empirical parameters *(11)*. Fragment-based methodology offers the possibility to extend such methods to larger systems, by limiting the length scale at which electron correlation effects are described at the highest level. Two rather different platforms for performing such calculations were described here, both united by the common goal of achieving an accurate description of non-covalent interactions in large clusters by dividing them up into smaller subsystems.

We have shown that a two-body expansion based on overlapping fragments (which requires independent electronic structure calculations on individual fragments, dimers of fragments, and intersections thereof) is capable of reproducing benchmark supersystem energies essentially exactly *(37,38)*, even for challenging examples such as fluoride–water clusters and in clusters as large as $(H_2O)_{57}$, the largest system that we have attempted to date. The downside of this approach is that it generates a rather large number of small electronic structure calculations, so that the number of independent calculations for a given supersystem size is much larger than that required for a traditional MBE *(38)*. Nevertheless, these small calculations are entirely independent and therefore embarrassingly parallelizable.

This approach holds great promise not only for extending quantum chemistry calculations to large cluster sizes but also for extending them to large basis sets, in order to perform many-body counterpoise corrections *(39,103,104)* and complete-basis extrapolations *(103,104)*. Indeed, with resolution-of-identity approaches *(105)*, the bottleneck step in obtaining MP2/CBS results is often the HF/aQZ calculation, and linear dependencies can severely hamper SCF convergence when large basis sets are used in medium-sized systems. The GMBE offers a path forward, and we are presently using this methodology to obtain MP2/CBS-quality results in systems where a supersystem calculation in the aQZ basis set would be difficult or impossible *(103,104)*. A variety of low-cost counterpoise correction schemes also emerge from applying the MBE to monomers and subclusters described using the cluster basis set *(39,103,104)*. Preliminary evidence indicates that the GMBE is relatively robust with respect to the details of the fragmentation procedure *(38)*, which also makes this a promising approach for use in macro-molecular systems, where fragmentation across covalent bonds is necessary. Such extensions are currently underway in our group.

The XSAPT method represents a different strategy, based not on combinatorics but on the physics of intermolecular interactions. The basic XSAPT idea *(51,52)* establishes a systematically improvable hierarchy of methods for describing intermolecular interactions in large systems, while the first rung of this hierarchy (a many-body generalization of SAPT0 *(15)*) achieves ~ 1 kcal/mol accuracy in intermolecular interaction energies, when combined with a suitable basis set *(51)*. (This version of XSAPT is due to appear in version 4.1 of the Q-Chem software package *(106,107)*.) However, results are rather sensitive to the choice of basis set *(52)*, indicating that error cancellation plays a key role in the success of the method.

One way to move beyond this undesirable situation is to climb further up the SAPT hierarchy, but a less expensive approach is to replace the problematic dispersion and exchange-dispersion terms in SAPT with empirical potentials *(53,54,89)*. Our latest version of this method, which we call XSAPT(KS)+D2 *(54)*, achieves a mean accuracy of <0.4 kcal/mol as compared to CCSD(T)/CBS benchmark dimer binding energies, performs well across the potential energy surfaces of challenging systems such as $F^-(H_2O)$, and exhibits cubic-scaling cost, both with respect to monomer size and with respect to the number of monomer units *(54)*. Unlike our first-generation XSAPT(KS)+D method *(53)*, the second-generation method is accurate not just for the total binding energies but also for the individual energy components such as dispersion, exchange-repulsion, and induction. As such, it is possible to generalize the SAPT energy decomposition to many-body systems *(54)*, to understand not just *what* the binding energy is, but *why* it has that value. Applications to understanding ligand binding in biomolecular systems are currently underway in our group.

ACKNOWLEDGEMENTS

Preliminary work on this project (papers of L. D. J.) was supported by a National Science Foundation CAREER Award (CHE-0748448 to J. M. H.) and by a Presidential Fellowship from The Ohio State University (to L. D. J.). The work of R. M. R. and K. U. L. is currently supported by the U.S. Department of Energy, Office of Basic Energy Sciences, Division of Chemical Sciences, Biosciences, and Geosciences, under Award No. DE-SC0008550 (to J. M. H.). Additional support has been provided by an Alfred P. Sloan Foundation Fellowship and a Camille Dreyfus Teacher-Scholar Award. Calculations were performed at the Ohio Supercomputer Center under project nos. PAS-0291 and PAA-0003.

REFERENCES

1. VandeVondele, J.; Borstnik, U.; Hutter, J. *J. Chem. Theory Comput.* **2012**, *8*, 3565–3573.
2. Yoo, S.; Aprà, E.; Zeng, X. C.; Xantheas, S. S. *J. Phys. Chem. Lett.* **2010**, *1*, 3122–3127.
3. Colbourn, E. A.; Douglas, A. E. *J. Chem. Phys.* **1976**, *65*, 1741–1745.
4. Yakovchuk, P.; Protozanova, E.; Frank-Kamenetskii, M. D. *Nucleic Acids Res.* **2006**, *34*, 564–574.
5. Autumn, K.; Sitti, M.; Liang, Y. A.; Peattie, A. M.; Hansen, W. R.; Sponberg, S.; Kenny, T. W.; Fearing, R.; Isrealachvili, J. N.; Full, R. J. *Proc. Natl. Acad. Sci. U. S. A* **2002**, *99*, 12252–12256.
6. McMullan, R. K.; Kvick, Å.; Popelier, P. *Acta Crystallgr. B* **1992**, *48*, 726–731.
7. Headen, T. F.; Howard, C. A.; Skipper, N. T.; Wilkinson, M. A.; Bowron, D. T.; Soper, A. K. *J. Am. Chem. Soc.* **2010**, *132*, 5735–5742.
8. Cohen, A. J.; Mori-Sanchez, P.; Yang, W. *Chem. Rev.* **2012**, *112*, 289–320.
9. Grimme, S. *WIREs Comput. Mol. Sci.* **2011**, *1*, 211–228.
10. Hopkins, B. W.; Tschumper, G. S. *J. Phys. Chem. A* **2004**, *108*, 2941–2948.
11. Hobza, P. *Acc. Chem. Res.* **2012**, *45*, 663–672.
12. Gordon, M. S.; Fedorov, D. G.; Pruitt, S. R.; Slipchenko, L. V. *Chem. Rev.* **2011**, *112*, 632–672.
13. Wen, S.; Nanda, K.; Huang, Y.; Beran, G. J. O. *Phys. Chem. Chem. Phys.* **2012**, *14*, 7579–7590.
14. Jeziorski, B.; Moszynski, R.; Szalewicz, K. *Chem. Rev.* **1994**, *94*, 1887–1930.
15. Hohenstein, E. G.; Sherrill, C. D. *WIREs Comput. Mol. Sci.* **2012**, *2*, 304–326.
16. Szalewicz, K. *WIREs Comput. Mol. Sci.* **2012**, *2*, 254–272.
17. Sebetci, A.; Beran, G. J. O. *J. Chem. Theory Comput.* **2010**, *6*, 155–167.
18. Vreven, T.; Morokuma, K. *Annu. Rep. Comput. Chem.* **2006**, *2*, 35–51.
19. Beran, G. J. O. *J. Chem. Phys.* **2009**, *130*(164115), 1–9.
20. Beran, G. J. O.; Nanda, K. *J. Phys. Chem. Lett.* **2010**, *1*, 3480–3487.
21. Nanda, K. D.; Beran, G. J. O. *J. Chem. Phys.* **2012**, *137*(174106), 1–11.
22. Mayhall, N. J.; Raghavachari, K. *J. Chem. Theory Comput.* **2011**, *7*, 1336–1343.
23. Christie, R. A.; Jordan, K. D. *Intermolecular Forces and Clusters II*; In: Wales, D., Christie, R.A., Eds.; Structure and Bonding, Vol. 116, Springer: Berlin, 2005; pp 27–41.
24. Cui, J.; Liu, H.; Jordan, K. D. *J. Phys. Chem. B* **2006**, *110*, 18872–18878.
25. Dahlke, E. E.; Truhlar, D. G. *J. Chem. Theory Comput.* **2007**, *3*, 46–53.
26. Dahlke, E. E.; Truhlar, D. G. *J. Chem. Theory Comput.* **2007**, *3*, 1342–1348.
27. Dahlke, E. E.; Truhlar, D. G. *J. Chem. Theory Comput.* **2008**, *4*, 1–6.
28. Dahlke, E. E.; Leverentz, H. R.; Truhlar, D. G. *J. Chem. Theory Comput.* **2008**, *4*, 33–41.
29. Ganesh, V.; Dongare, R. K.; Balanarayan, P.; Gadre, S. R. *J. Chem. Phys.* **2006**, *125* (104109), 1–10.

30. Rahalkar, A. P.; Ganesh, V.; Gadre, S. R. *J. Chem. Phys.* **2008**, *129*(234101), 1–7.
31. Rahalkar, A. P.; Katouda, M.; Gadre, S. R.; Nagase, S. *J. Comput. Chem.* **2010**, *31*, 2405–2418.
32. Sahu, N.; Yeole, S. D.; Gadre, S. R. *J. Chem. Phys.* **2013**, *138*(104101), 1–6.
33. Li, W.; Li, S.; Jiang, Y. *J. Phys. Chem. A* **2007**, *111*, 2193–2199.
34. Hua, S.; Hua, W.; Li, S. *J. Phys. Chem. A* **2010**, *114*, 8126–8134.
35. Li, W.; Hua, W.; Fang, T.; Li, S. In: *Computational Methods for Large Systems: Electronic Structure Approaches for Biotechnology and Nanotechnology*; Reimers, J.R., Ed.; Wiley: Hoboken, NJ, 2011; pp 227–258.
36. Mayhall, N. J.; Raghavachari, K. *J. Chem. Theory Comput.* **2012**, *8*, 2669–2675.
37. Richard, R. M.; Herbert, J. M. *J. Chem. Phys.* **2012**, *137*(064113), 1–17.
38. Richard, R. M.; Herbert, J. M. *J. Chem. Theory Comput.* **2013**, *9*, 1408–1416.
39. Kamiya, M.; Hirata, S.; Valiev, M. *J. Chem. Phys.* **2008**, *128*(074103), 1–11.
40. Fedorov, D. G.; Nagata, T.; Kitaura, K. *Phys. Chem. Chem. Phys.* **2012**, *14*, 7562–7577.
41. Tschumper, G. S. *Chem. Phys. Lett.* **2006**, *427*, 185–191.
42. Elsohly, A. M.; Shaw, C. L.; Guice, M. E.; Smith, B. D.; Tschumper, G. S. *Mol. Phys.* **2007**, *105*, 2777–2782.
43. Bates, D. M.; Smith, J. R.; Tschumper, G. S. *J. Chem. Theory Comput.* **2011**, 7, 2753–2760.
44. Chałasiński, G.; Szczęśniak, M. M.; Kendall, R. A. *J. Chem. Phys.* **1994**, *101*, 8860–8869.
45. Turki, N.; Milet, A.; Rahmouni, A.; Ouamerali, O.; Moszynski, R.; Kochanski, E.; Wormer, P. E. S. *J. Chem. Phys.* **1998**, *109*, 7157–7168.
46. Pedulla, J. M.; Kim, K.; Jordan, K. D. *Chem. Phys. Lett.* **1998**, *291*, 78–84.
47. Tauer, T. P.; Sherrill, C. D. *J. Phys. Chem. A* **2005**, *109*, 10475–10478.
48. Ringer, A. L.; Sherrill, C. D. *Chem. Eur. J.* **2008**, *14*, 2452–2547.
49. Chen, Y.; Li, H. *J. Phys. Chem. A* **2010**, *114*, 11719–11724.
50. Xie, W.; Song, L.; Truhlar, D. G.; Gao, J. *J. Chem. Phys.* **2008**, *128*(234108), 1–9.
51. Jacobson, L. D.; Herbert, J. M. *J. Chem. Phys.* **2011**, *134*(094118), 1–17.
52. Herbert, J. M.; Jacobson, L. D.; Lao, K. U.; Rohrdanz, M. A. *Phys. Chem. Chem. Phys.* **2012**, *14*, 7679–7699.
53. Lao, K. U.; Herbert, J. M. *J. Phys. Chem. Lett.* **2012**, *3*, 3241–3248.
54. Lao, K. U.; Herbert, J. M. *J. Chem. Phys.* **2013**, *139*(034107), 1–16.
55. Deev, V.; Collins, M. A. *J. Chem. Phys.* **2005**, *122*(154102), 1–12.
56. Kohn, W. *Phys. Rev. Lett.* **1996**, *76*, 3168–3171.
57. Prodan, E.; Kohn, W. *Proc. Natl. Acad. Sci. U. S. A.* **2005**, *102*, 11635–11638.
58. Collins, M. A.; Deev, V. A. *J. Chem. Phys.* **2006**, *125*(104104), 1–15.
59. Addicoat, M. A.; Collins, M. A. *J. Chem. Phys.* **2009**, *131*(104103), 1–9.
60. Collins, M. A. *Phys. Chem. Chem. Phys.* **2012**, *14*, 7744–7751.
61. Leverentz, H. R.; Truhlar, D. G. *J. Chem. Theory Comput.* **2009**, *5*, 1573–1584.
62. Lao, K. U.; Herbert, J. M. *J. Phys. Chem. A* **2012**, *116*, 3042–3047.
63. Das, D.; Eurenius, K. P.; Billings, E. M.; Sherwood, P.; Chatfield, D. C.; Hodošček, M.; Brooks, B. R. *J. Chem. Phys.* **2002**, *117*, 10534–10547.
64. Xie, W.; Gao, J. *J. Chem. Theory Comput.* **2007**, *3*, 1890–1900.
65. Xie, W.; Orozco, M.; Truhlar, D. G.; Gao, J. *J. Chem. Phys.* **2009**, *5*, 459–467.
66. Gao, J. *J. Chem. Phys.* **1998**, *109*, 2346–2354.
67. Jurečka, P.; Šponer, J.; Černý, J.; Hobza, P. *Phys. Chem. Chem. Phys.* **2006**, *8*, 1985–1993.
68. Takatani, T.; Hohenstein, E. G.; Malagoli, M.; Marshall, M. S.; Sherrill, C. D. *J. Chem. Phys.* **2010**, *132*(144104), 1–5.
69. Han, J.; Truhlar, D. G.; Gao, J. *Theor. Chem. Acc.* **2012**, *131*(1161), 1–15.
70. Khaliullin, R.; Head-Gordon, M.; Bell, A. T. *J. Chem. Phys.* **2006**, *124*(204105), 1–11.

71. Breneman, C. M.; Wiberg, K. B. *J. Comput. Chem.* **1990**, *11*, 361–373.
72. Lotrich, V. F.; Szalewicz, K. *J. Chem. Phys.* **1997**, *106*, 9668–9687.
73. Lotrich, V. F.; Szalewicz, K. *J. Chem. Phys.* **2000**, *112*, 112–121.
74. Moszynski, R.; Wormer, P. E. S.; Jeziorski, B.; van der Avoird, A. *J. Chem. Phys.* **1995**, *103*, 8058.
75. Williams, H. L.; Chabalowski, C. F. *J. Phys. Chem. A* **2001**, *105*, 646–659.
76. Misquitta, A. J.; Szalewicz, K. *Chem. Phys. Lett.* **2002**, *357*, 301–306.
77. Pople, J. A.; Krishnan, R.; Schegel, H. B.; Binkley, J. S. *Int. J. Quantum Chem. Symp.* **1979**, *13*, 225–241.
78. Jeziorski, B.; Moszynski, R.; Ratkiewicz, A.; Rybak, S.; Szalewicz, K.; Williams, H. L. In: Clementi, E., Ed.; *Methods and Techniques in Computational Chemistry: METECC-94*; STEF: Cagliari, 1993; Vol. B, pp 79–129. (chapter 3).
79. Williams, H. L.; Mas, E. M.; Szalewicz, K.; Jeziorski, B. *J. Chem. Phys.* **1995**, *103*, 7374–7391.
80. Wolinski, K.; Pulay, P. *J. Chem. Phys.* **2003**, *118*, 9497–9503.
81. Steele, R. P.; DiStasio, R. A., Jr.; Shao, Y.; Kong, J.; Head-Gordon, M. *J. Chem. Phys.* **2006**, *125*(074108), 1–11.
82. Rohrdanz, M. A.; Martins, K. M.; Herbert, J. M. *J. Chem. Phys.* **2009**, *130*(054112), 1–8.
83. Burnham, C. J.; Xantheas, S. S. *J. Chem. Phys.* **2002**, *116*, 1479–1492.
84. Hohenstein, E. G.; Sherrill, C. D. *J. Chem. Phys.* **2010**, *133*(014101), 1–12.
85. Iikura, H.; Tsuneda, T.; Yanai, T.; Hirao, K. *J. Chem. Phys.* **2001**, *115*, 3540–3544.
86. Henderson, T. M.; Janesko, B. G.; Scuseria, G. E. *J. Phys. Chem. A* **2008**, *112*, 12530–12542.
87. Lange, A. W.; Rohrdanz, M. A.; Herbert, J. M. *J. Phys. Chem. B* **2008**, *112*, 6304–6308.
88. Baer, R.; Livshits, E.; Salzner, U. *Annu. Rev. Phys. Chem.* **2010**, *61*, 85–109.
89. Hesselmann, A. *J. Phys. Chem. A* **2011**, *115*, 11321–11330.
90. Grimme, S. *J. Comput. Chem.* **2006**, *27*, 1787–1799.
91. Tozer, D. J.; Handy, N. C. *J. Chem. Phys.* **1998**, *109*, 10180–10189.
92. Gruning, M.; Gritsenko, O. V.; van Gisbergen, S. J. A.; Baerends, E. J. *J. Chem. Phys.* **2001**, *114*, 652–660.
93. Řezáč, J.; Riley, K. E.; Hobza, P. *J. Chem. Theory Comput.* **2011**, 7, 2427–2438.
94. Hohenstein, E. G.; Sherrill, C. D. *J. Chem. Phys.* **2010**, *132*(184111), 1–10.
95. DiStasio, R. A., Jr.; Head-Gordon, M. *Mol. Phys.* **2007**, *105*, 1073–1083.
96. Pitoňák, M.; Řezáč, J.; Hobza, P. *Phys. Chem. Chem. Phys.* **2010**, *12*, 9611–9614.
97. Pitoňák, M.; Neogrády, P.; Černý, J.; Grimme, S.; Hobza, P. *ChemPhysChem* **2009**, *10*, 282–289.
98. Gordon, M. S.; Freitag, M. A.; Bandyopadhyay, P.; Jensen, J. H.; Kairys, V.; Stevens, W. J. *J. Phys. Chem. A* **2001**, *105*, 293–307.
99. Gordon, M. S.; Slipchenko, L. V.; Li, H.; Jensen, J. H. *Annu. Rep. Comput. Chem.* **2007**, *3*, 177–193.
100. Flick, J. C.; Kosenkov, D.; Hohenstein, E. G.; Sherrill, C. D.; Slipchenko, L. V. *J. Chem. Theory Comput.* **2012**, *8*, 2835–2843.
101. Herbert, J. M.; Head-Gordon, M. *J. Am. Chem. Soc.* **2006**, *128*, 13932–13939.
102. Podeszwa, R.; Pernal, K.; Patkowski, K.; Szalewicz, K. *J. Phys. Chem. Lett.* **2010**, *1*, 550–555.
103. Richard, R. M.; Lao, K. U.; Herbert, J. M. *J. Phys. Chem. Lett.* **2013**, *4*, 2674–2680.
104. Richard, R. M.; Lao, K. U.; Herbert, J. M. Manuscript in preparation.
105. Sherrill, C. D. *J. Chem. Phys.* **2010**, *132*(110902), 1–7.
106. Shao, Y.; et al. *Phys. Chem. Chem. Phys.* **2006**, *8*, 3172–3191.
107. Krylov, A. I.; Gill, P. M. W. *WIREs Comput. Mol. Sci.* **2013**, *3*, 317–326.
108. Misquitta, A. J.; Szalewicz, K. *J. Chem. Phys.* **2005**, *122*(214109), 1–19.

Bioinformatics

Section Editor: Wei Wang

Department of Chemistry and Biochemistry,
University of California-San Diego, La Jolla, CA, USA

CHAPTER THREE

Atomistic Modelling of Phosphopeptide Recognition for Modular Domains

Chia-en A. Chang[1], Yu-ming M. Huang

Department of Chemistry, University of California, Riverside, California, USA
[1]Corresponding author: e-mail address: chiaenc@ucr.edu

Contents

Abstract

This review aims at discussing the molecular details of binding specificity, promiscuity and mechanisms of phosphopeptide recognition to modular domains using computational tools. Protein–phosphoprotein interactions are the driving forces that underline multiple signalling events which are important in cellular function. Understanding protein–phosphopeptide recognition assists designing phosphopeptide sequences as inhibitors to manipulate protein–protein interactions for cell biology studies and therapeutics. Notably, the modular domain–phosphopeptide binding is mostly promiscuous and weak binding, which significantly differs from most protein–ligand binding systems used for drug design. In this chapter, we review recent advances in computational work for modular domain–phosphopeptide binding and knowledge gained for their binding mechanisms. We discuss the phosphopeptide binding of modular domains in DNA damage responses, FHA and BRCT, the smallest modular domain WW and the tyrosine kinase binding domain, such as Cbl–TKB that recognizes

Annual Reports in Computational Chemistry, Volume 9
ISSN 1574-1400
http://dx.doi.org/10.1016/B978-0-444-62672-1.00003-0

61

phosphotyrosine. We also discuss challenges and possible future directions for improving peptide design as inhibitors for phosphoprotein–protein binding.

1. INTRODUCTION

Protein function must be regulated and protein phosphorylation is one of the major tools that have developed to fill this need. Protein phosphorylation is widely exploited in DNA damage repair, signal transduction, cell growth and cell cycle regulation; the cascades of downstream signals can be triggered by grabbing a certain phosphoprotein *(1–6)*. Diverse arrays of modular domains recognize phosphoproteins by non-covalent binding that result in switch-like changes in protein function. The interactions between modular domains and their phosphoprotein partners are usually weak and transient, which tend to be underrepresented in high-throughput and computational studies. The domains are usually promiscuous proteins and can recognize more than one peptide sequence. Understanding these protein–peptide interactions is of great interest in a wide variety of applications such as molecular detection, inhibitor discovery and searching for binding partners. Promiscuous recognition at the molecular level usually involves different levels of conformational changes and numerous bond rotations *(7,8)*. However, the underlying mechanism that drives diverse ligands to dock into the same binding site of a protein is not fully understood.

The phosphoproteins are usually classified into two families: phosphotyrosine (pTyr)-containing and phosphoserine (pSer)/phosphothreonine (pThr)-containing sequences, which are phosphorylated and dephosphorylated by different categories of kinases (e.g. pTyr kinase and pThr/pSer kinase) and phosphatases *(9)*. Recent studies discovered a few modular domains that particularly recognize pThr/pSer- or pThr-containing sequences, such as the breast cancer-associated protein BRCA1 C-terminal (BRCT) repeats, WW domain, 14-3-3, WD40, polo-box and forkhead-associated (FHA) domain *(10–12)*. Domains such as the Src homology-2 (SH2) domain and phosphotyrosine kinase binding (TKB) domain mediate protein–protein interactions by targeting motifs that contain pTyr *(8,13,14)*.

The recent development of innovative algorithms has advanced knowledge of molecular recognition in phosphopeptide–modular domain systems. For example, advances in molecular dynamics (MD) simulations, various post-analysis methods and bioinformatics algorithms offer powerful tools to study binding mechanisms (Figure 3.1), assist computer-based inhibitor design and predict phosphopeptide binding sites *(15–22)*. In this chapter, we discuss the mechanisms of binding various phosphopeptides to modular domains. Specifically, we ask how nature achieves both specificity and promiscuity in these important signal transduction systems.

2. METHODS

2.1. Atomistic molecular dynamics simulations

The classical MD simulation-based method is the most widely used tool of computational techniques. It is perhaps the least glamorous method, but in many cases, it is the most powerful method. It provides atomistic details of motions of a molecular system and inter- and intra-molecular interactions during a given simulation time period. MD involves the integration of Newton's equation of motion for a set of atoms. The solvent can be modelled either explicitly, which is more accurate but more expensive, or implicitly through use of approximations based on continuum electrostatics, which is less accurate and inexpensive. Molecular mechanics (MM) force fields such as the OPLS, AMBER and CHARMM are typically used in these simulations with fixed-charged models *(23–26)*. Parameters for phosphoresidues, such as pThr, pSer and pTyr, are also available *(27)*. Conventional MD simulations can be used in an unbiased manner, but they can only realistically probe nanosecond-order timescales. Other biased sampling methods, such as targeted MD, accelerated MD and torsion angle-based sampling methods can be used to explore conformational changes more efficiently or to compute binding energies *(28–31)*. MM methods cannot be used to study catalysis, and combined quantum mechanics (QM) and MM have been used to study the phosphorylation reaction and how phosphorylation may affect catalysis. Because our focus is non-covalent binding and phosphopeptide recognition, we do not discuss QM/MM methods.

2.2. Bioinformatics tools

A modular domain may be identified in many different proteins, and although their sequences may not be highly identical, the domains share the same structure features. For example, the FHA domain has been identified in more than 2000 proteins from the Pfam database *(32,33)*, and the structure is conserved. Moreover, modular domains may form complexes with several different phosphopeptide sequences and the complex

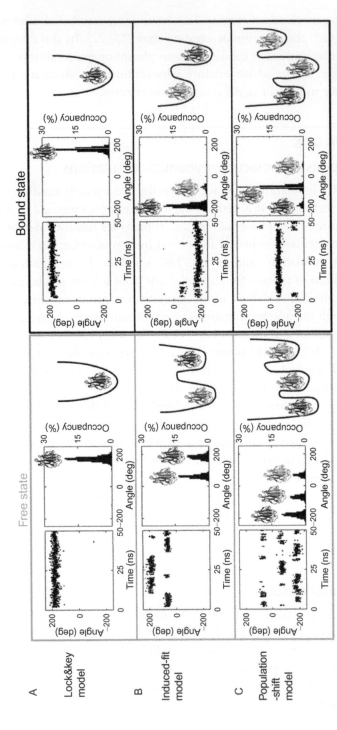

Figure 3.1 Three hypothesis models, lock and key (A), induced-fit (B) and population-shift/conformation-selection (C), to represent conformational changes in free and bound state. We plot the specific dihedral angle with simulation time and make it as a histogram distribution. The proteins coloured as blue, red and green indicate different conformations. (For interpretation of the references to colour in this figure legend, the reader is referred to the online version of this chapter.)

conformations may not be highly identical. X-ray crystallography and nuclear magnetic resonance (NMR) provide the highest resolution structural information for the domain–phosphopeptide complexes but are also the most difficult and time consuming to obtain. Bioinformatics methods provide fast and powerful tools to model the binding of various peptide sequences to a modular domain of a protein or different proteins. Homology modelling and multi-sequence alignment tools using well-developed Web sites are popular methods to build models for domains without experimental structures. The first step of homology modelling methods begins with the selection of suitable structural template(s) from the Protein Data Bank (PDB). Web servers such as SWISS-MODEL provide user friendly interface to search for templates *(34–36)*. The server also provides a template library, the SWISS-MODEL template Library (ExPDB), which is derived from the PDB. A wide variety of alignment tools and homology modelling packages and servers, such as T-coffee, MODELLER, Sybyl, Prime and ICM, are also commonly used to develop a homology model based on the selected template(s) *(37)*.

2.3. Coarse-grained Brownian dynamics simulations

The introduction of a phosphate group generally carries a -2 charge at physiologically pH. The large electrostatic perturbation can affect association processes of the phosphopeptide or phosphoprotein by increasing the electrostatic steering forces and/or alter protein structure and interactions. To study molecular association processes and large-scale conformational changes are being accomplished through the use of course-grained (CG) model and several CG models have been implemented in current MD or Brownian dynamics (BD) software *(38–41)*. Many CG models have been proposed over the years, for example, see refs. *(42–56)*. In this chapter, we discuss the use of BD algorithm with a one-bead CG model developed by Tozzini and McCammon as an example to investigate phosphopeptide–domain association *(57,58)*.

3. EXAMPLES OF MODELLING OF PHOSPHOPEPTIDE AND MODULAR DOMAIN BINDING

3.1. FHA domains: a domain that specifically recognizes phosphothreonine

A unique feature of FHA domains is that they can differentiate pThr-containing peptides from pSer-containing peptides, although the difference is only one methyl group *(59–63)*. Most pSer/pThr-binding modules, such as BRCT and WW domains, can specifically bind to short pSer/pThr-containing motifs equally well. Not unsurprisingly, one more methyl group of Thr has few effects on the overall binding and molecular recognition. Not only in peptide–protein binding, a recent literature analyses more than 2000 cases of methyl effects on protein-drug like compound and the results shows that less than 10% cases, containing a methyl group, results in 10-fold or greater binding affinity boost *(64)*. However, most FHA domains recognize only pThr residue in target proteins, and the substitution of pSer for pThr in model peptides severely weakens binding. Because Ser/Thr kinase phosphorylates both residues, the FHA domain can efficiently reduce potential interaction sites by specifically binding to pThr-containing regions and this recognition may function as a filter to further select the protein partner.

3.1.1 Topological and 3D Structures of FHA

The FHA domain is associated with proteins of diverse functions in different organisms. For example, the FHA1 domain in Rad53 protein, the Rad53–FHA1 domain, interacts with phosphorylated Rad9 in response to DNA damage, and the Dun1–FHA domain interacts with SCD1 of Rad53, which leads to activation of Dun1 in response to DNA damage response *(65,66)*. Phosphopeptide sequences discussed in this chapter are listed in Table 3.1. Although the sequence identity between the FHA from different proteins is low, for example, the sequence identities of Dun1–FHA and the Ki67–FHA to Rad53–FHA1 are both 34%, the structure of the FHA domain is well conserved: a twisted β sandwich of 11 well-defined β sheets, 5 in the front and 6 at the back (Figure 3.2) *(61,62,67,75–79)*. Generally, the domain contains ~120–140 residues, but only 5–10 residues are conserved. Six loops connected to the secondary β strands constructing the pThr-binding site are the main difference between distinct FHA domains. Experimental structures show that the synthetic peptides bind to the loops between β3–β4, β4–β5, β6–β7 and β10–β11 and the conserved pThr-binding site locates between loops β4–β5 and β6–β7 (Figure 3.2) *(61,62)*.

3.1.2 Revealing the Specificity of the pThr-Binding Site in FHA Domain

A unique feature of the FHA domain family is the use of loops to recognize peptide sequences. Loops are typically considered flexible regions in proteins, but analysis of FHA domain conformations suggests a conserved

Table 3.1 List of phosphopeptide sequences

No.	PDB ID	Domain	Method	K_d (μM)	Sequence	References
1	1G6G	Rad53-FHA1	X-ray	0.53	LEV(pT)EADATFAK	(61)
2	1K3Q	Rad53-FHA1	NMR	0.30	SLEV(pT)EADATFVQ	(65)
3	2JQL	Dun1-FHA	NMR	0.30–1.2	NI(pT)QP(pT)QQST	(66)
4	2AFF	Ki67-FHA	NMR	0.077	KTVD(pS)QGP(pT)PVC(pT)**PTFLERRKSQVAELN**DDKDD**EIVFK**QPISC	(67)
5	1F8A	Pin1-WW	X-ray	34	Y(pS)PT(pS)PS	(68)
6	1T2V	BRCA1–BRCT	X-ray	0.40	AAYDI(pS)QVFPFA	(69)
7	1T29	BRCA1–BRCT	X-ray	0.90	ISRST(pS)PTFNKQ	(70)
8	1Y98	BRCA1–BRCT	X-ray	3.7	PTRVS(pS)PVFGA	(71)
9	3COJ	BRCA1–BRCT	X-ray	5.2	PQ(pS)PTFPEAG	(72)
10	2CBL	Cbl-TKB	X-ray	12.2	TLNSDGpYTPEPA	(73)
11		Cbl-TKB	X-ray[a]	3.7	pYTPEP	(74)
12		Cbl-TKB		>37[b]	pYTP(ptE)P	

Secondary structure, α-helix and β-sheet are labelled as bold.
[a]Manually truncated the long peptide from 2CBL.
[b]Unpublished data.

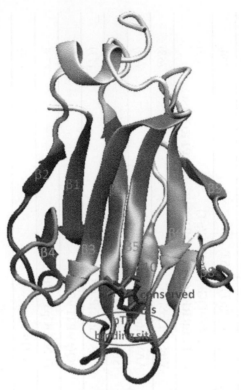

Figure 3.2 FHA structure from a MD snapshot. Loops/β-sheets in the front and at the back are labelled as yellow and blue, respectively. Other loops which do not have direct contacts with peptide are shown in white. Green and red represent conserved His- and pThr-binding cavity. (For interpretation of the references to colour in this figure legend, the reader is referred to the online version of this chapter.)

structure in the main pThr-binding site formed by two loops between β3–β 4 and β6–β7 (see Figure 3.2). Recent studies substituted pThr with pSer *in silico* and performed interaction energy calculations and analysis of configuration space to illustrate why the methyl group plays the crucial role in determining binding *(80)*.

Although experimental structures demonstrate a pocket to accept the pThr methyl group, the static conformation cannot ascertain that pSer fails to form equally good interactions with the nearby residues because the protein is dynamic and may fill the space of a methyl group by slightly changing the protein conformations. MD simulations show that the cavity is highly suited to pThr, and the local flexibility is changed because of the lack of the methyl group in pSer-containing phosphopeptides. Most side-chain dihedrals stay in the same rotamer states for both pSer/pThr residues, but the second pThr side-chain dihedral angle (see Figure 3.3A) differs. Figure 3.3 shows the distribution of the side-chain dihedral angle of pThr and pSer in Rad53–FHA1 peptide. The dihedral of pSer deviates from pThr with a significant angle shift and also has wider distribution, so the dihedral is more flexible. Clearly, the methyl group of pThr allows the phosphoresidue to fill the entire pocket of the binding site, and no room is available for spacious vibration of the pThr side chain. In contrast, the space released by the absence of methyl group cannot be adequately filled by protein side chains, which creates room for the dihedral of pSer to be more flexible. Although side chains of the dual loops are mobile, the conserved His, located at the N-terminus of β5, uses the imidazole ring and polar interactions to form stable interactions with residues of loops β4–β5 and β6–β7. Note that the His shown in Figure 3.2 is conserved in all FHA domains, although it does not directly interact with a phosphopeptide. Interestingly, MD simulations explain how FHA makes use of the conserved His to stabilize the dual loop and form a structural room to dock the methyl group and discriminate pThr/pSer.

Interaction energy calculations, so-called MM/PBSA calculations, quantify the attraction forces between pSer/pThr and residues around the phosphoresidue, which reveal how FHA can discriminate between them. Although the only difference between the Thr and Ser residue is one methyl group, which is usually considered not significant, atomistic modelling indicates that the methyl group directly interacts with residues of loops β4–β5 and β6–β7 of the FHA domain (see Figure 3.3). The local interaction energy is less favourable when pThr is replaced by pSer and shows that van der Waals interactions are weakened considerably by the lack of a single methyl group of pSer; the loss of the van der Waals attraction can be weakened by ~3 kcal/mol. The interaction between the methyl group of pThr and the nearby residues is unlikely to be 3 kcal/mol, but instead, the computed energy reveals the crucial role of the methyl group to stabilize the complex conformation locally. Interestingly, although the phosphate group of pSer still retains hydrogen bonding between the nearby residues of FHA, the electrostatic attractions are still weakened. This again supports that solely forming H-bonds between the phosphate group of the phosphoresidue is not enough for phosphopeptide and FHA domain binding, and lacking the methyl group destabilizes the complex. Fewer contacts can be formed when pSer is present in the peptide.

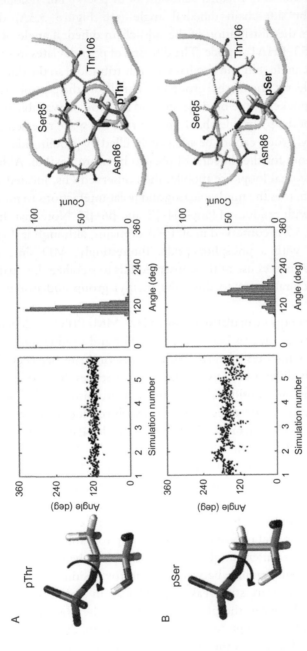

Figure 3.3 The second side-chain dihedral angle of pThr (A) and pSer (B). We plot the angle distribution from five individual MD simulations; each simulation is 1 ns. The MD snapshots show the detailed interactions between phosphoresidue and FHA domain. Blue dash lines indicate charge–charge attractions. (For interpretation of the references to colour in this figure legend, the reader is referred to the online version of this chapter.)

3.2. WW: The smallest modular domain recognizes both pThr and pSer

WW domains are the smallest, naturally occurring modular domains that have only about 40 residues *(4)*. Due to the small size and compact fold, WW became an attractive model for studying protein stability, protein design and prediction of binding phosphopeptides for a specific WW domain sequence *(68,81)*. They are found in many different proteins; for example, the WW domain in Pin1, Pin1-WW, is essential for mitotic progression *(82)*.

3.2.1 Topological and 3D structures of WW

The name refers to two tryptophan (W) residues that are 20–22 residues apart and are present in most WW domains in sequence analysis. Not all WW domains function as phosphopeptide binding modules, but a subgroup of the domain recognizes proline (Pro)-containing phosphopeptides and folds into three anti-parallel β stands. As shown in Figure 3.4, Arg21 and Ser22 residues in the loops between β1 and β2 are the phosphate group recognition sites *(4,5,83)*. It specifically binds to pThr-Pro- or pSer-Pro-containing motifs with slightly higher affinity for pThr-Pro-containing peptides *(81)*. For example, in the Pin1-WW domain, the aromatic rings of Tyr23 and Trp34 define a steric clamp to confer a Pro adjacent to pSer/pThr *(83)*.

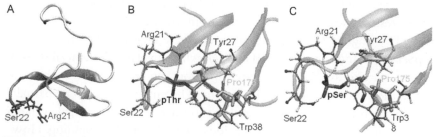

Figure 3.4 (A) WW domain structure. Yellow indicates peptide binding loop. Two key residues which directly bind to phosphoresidue are represented in blue and green. The MD snapshot that shows pThr and pSer-containing peptide binding is shown as (B) and (C), respectively. (For interpretation of the references to colour in this figure legend, the reader is referred to the online version of this chapter.)

3.2.2 Intrinsic dynamics of apo WW domain and pThr and pSer recognition

Flexibility helps accommodate the promiscuous binding properties of modular domains (1, 12). For domains that show obvious two-state—the free and bound—conformations, such as Pin1-WW, the free domains may pre-exist holo-like conformations. NMR relaxation experiments have emerged as a premier tool for revealing the location and timescale of conformation ensembles for free domains and the transitions between the free and bound states (84–86). However, detailed atomistic observations still need to rely on computational simulations. MD simulations and Markov state models (MSMs) show that the apo Pin1-WW domain has intrinsic dynamics, resulting in conformational equilibria between holo-like and pre-exist alternative conformers for phosphopeptide recognition. The analysis of correlated motions also suggests that the domain may help couple the substrate binding site on the WW domain to the one on the catalytic domain (87).

Although it is common that proteins treat Ser and Thr as a similar residue, knowing how WW has non-specific pThr/pSer recognition with a specific Pro residue is of interest. The WW domain utilizes the combination of a single loop with β-sheet, which allows effectively side-chain rearrangement to accept both pSer and pThr. Although the WW domains are able to recognize both pSer- and pThr-containing peptides, binding energy calculations from MD simulations suggest that the domain favours pThr because of the more preferable van der Waals attractions. This trend is in agreement with experimental results (81, 82). As shown in Figure 3.4B and C, two conserved aromatic residues of the domain, Tyr and Trp, create a cavity, but no side chains of phosphopeptides could nicely fit into the cavity during MD simulations. Interestingly, the conserved Pro residue adjacent to the phosphoresidue is clamped by Tyr and Trp, which stays in the cavity and further restricts nearby phosphopeptide conformations. The confined region formed by rings of Tyr, Trp and Pro is conserved regardless of the presence of pSer or pThr (see Figure 3.4B and C), which explains the crucial roles of Pro. Because of the bulky ring conformations, an empty space is observed during the course of the MD simulations. The empty space can be partially filled by the methyl group of pThr, thus resulting in more favourable van der Waals interactions and a less flexible side chain while pThr is binding. However, the Pro residue but not the methyl group of pThr primarily occupies the cavity in phosphopeptide recognition. Therefore, the domains do not show significant discrimination between pSer and pThr. The hydrogen bonds between pThr or pSer and the nearby residues are always present during the simulations, suggesting the important role of phosphate for over-all binding affinities.

3.3. BRCT: Case study for binding mechanisms

Promiscuous proteins are commonly observed in biological systems, and this characteristic makes it possible to use limited numbers of domains to regulate considerably more binding partners through protein phosphorylation. In addition to the ability to recognize phosphopeptides with similar polar or non-polar features, the binding interface of some domains can accommodate diverse peptide sequences, including both polar *and* non-polar phospho-peptides, by applying different binding mechanisms. For this chapter, we use the BRCT domain of BRCA1, BRCA1–BRCT, as a model system to study binding mechanism for promiscuous recognition (see Table 3.1 for phosphopeptide sequences).

Because of the promiscuous nature, a domain may utilize different bind-ing mechanisms when it recruits different phosphopeptide sequences. Typ-ically, the binding mechanisms may be divided into three representative postulates: lock and key, induced-fit and population-shift/conformational-selection model *(88,89)*. The basic mechanism of lock and key model has been introduced for a long time *(90)*. In this postulate, only correctly pre-organized ligands are capable to fit into the active site of the lock, which indicates that the conformations would not change during binding (Figure 3.1A) *(91–93)*. Since not all cases can be adequately explained by the rigid lock model, another assumption, induced-fit theory, has been pro-posed and showed that biomolecules are rather flexible structures in which the conformation can be reshaped and distorted to optimize interactions with partners (Figure 3.1B) *(88,89,91)*. Experiments also support another mechanism, termed population-shift or conformational-selection model. In this model, free protein or ligand has a large number of conformations pre-existing in the native state. After binding, the structure is perturbed and thus the properties of population switch (Figure 3.1C) *(88,94)*. Because the BRCA1–BRCT domain recognizes sequences with mainly polar, non-polar and mixed residues, it serves as a good model system to study binding mechanisms to gain knowledge in peptide design.

3.3.1 Topological and 3D structures of BRCT

The BRCT domains are protein-binding modules originally discovered as a domain conserved in multiple DNA damage-response proteins *(95,96)*. This vital modular domain system, BRCA1–BRCT, shows strong binding affinity with different types of pSer peptides *(69,70,97)*. The BRCT domains are often found in the form of tandem repeats (see Figure 3.5A). A single domain in the BRCT fold is packed with four parallel β sheets with a pair of α helices (α1 and α3) flanked on each side and one helix on the opposite face (Figure 3.5A) *(98)*. The overall structure of the tandem repeats BRCT domain from different protein families is conserved: a cluster of hydrophobic residues locates at the interface between the two repeats *(97)*. The mutation of the BRCT will cause cancers. Unlike the domain parts, the linker connecting two repeats is more diverse, with poorly defined crystal structures, which suggests flexibility of the linker region *(99)*. The BRCA1 protein is a tumour suppressor and is involved in multiple cellular functions such as DNA repair and signal transduction. Besides BRCA1, many other proteins are known to contain single or multiple BRCT domains, with 90–100 amino acid residues in each domain. In this chapter, we focus on BRCA1–BRCT. Similar to WW, not all BRCT domains function as phosphopeptide-binding modules. However, the BRCA1–BRCT domains recognize pSer-X-X-Phe motifs (X means several different amino acids) that have been demonstrated to bind the phosphorylated DNA helicase BACH1, phosphorylated Abraxas and the phosphorylated transcriptional corepressor CtIP *(11,70–72)*.

3.3.2 Mechanistic insights into BRCT–phosphopeptide recognition

MD simulations can efficiently sample dihedral rotations in both free and bound domain, which provides us the useful information about which mechanism that the interface will apply to recognize a given phosphopeptide sequence. The rotamer states of each residue in the binding site of BRCT and four different phosphopeptides can be counted from the MD simulations. Modelling results show that the backbone dihedral angle mostly stayed in a single energy well and remained the same with or without phosphopeptide binding, and as expected, the side-chain dihedrals were more flexible and usually had more than one rotamer state. Although, commonly, ligand binding decreases the number of available rotamer states, the rotamer states of BRCT and the phosphopeptides can be increased or remain unchanged in their free and bound states (Figure 3.5B). Upon binding to BRCA1–BRCT, phosphopeptide 6 showed new rotamer states of a number of residues shown in Figure 3.5C, which suggests that the peptide can efficiently rearrange itself to fit into the binding interface of BRCT. New rotamer states of several non-polar residues of phosphopeptide 6 are induced by binding a hydrophobic binding groove. This new dihedral angle

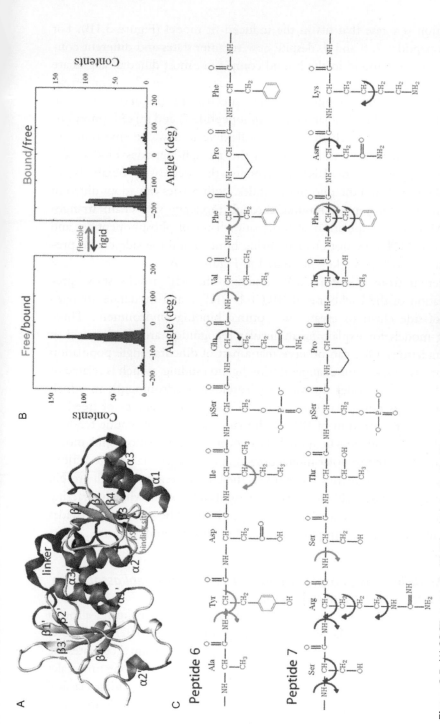

Figure 3.5 (A) BRCT structure. (B) Dihedral angle distribution to represent conformations can be either more flexible or rigid in bound state. (C) Two peptide sequences with all rotatable bonds. Red and blue show that, after binding, dihedrals become more flexible and rigid, respectively. (For interpretation of the references to colour in this figure legend, the reader is referred to the online version of this chapter.)

distribution is a case that fits in the induced-fit model (Figure 3.1B). For phosphopeptides 7, 8 and 9, despite new rotamer states and different conformations being found in the bound complexes, most dihedral angles are less mobile and have fewer rotamer states, which utilizes the "conformational selection" or "population-shift" mechanism (Figures 3.1C and 3.5C). Therefore, highly charged (phosphopeptide 7) and mixed (phosphopeptides 8 and 9) side chains are more flexible in a water environment, but the movement is restrained when the phosphopeptides bind to the hydrophobic groove of BRCT. Despite the decreased flexibility of BRCA1–BRCT and the phosphopeptides, both parties can adopt different conformations for peptide binding through moderate side-chain arrangement. Moreover, population of some side chains of phosphopeptides and BRCA1–BRCT does not change during binding and the side chains preorganize as a lock and a key illustrated in Figure 3.1A.

Different from the Pin1-WW domain, the MD results show preorganization of the backbone of BRCA1–BRCT and moderate arrangements of side chain to create the optimal binding environment. Three binding models for explaining molecular recognition are all observed in MD simulations. One can observe that a part of dihedral angle population does not have significant changes during ligand binding, which is related to the lock and key model. In more hydrophobic environments, in which binding enables non-polar side chains to have more degrees of freedom to rotate, the recognition of BRCT domain and phosphopeptide tends to proceed by the induced-fit model. However, the population-shift model is commonly found in polar interactions due to charge–charge interactions that have preferred direction. Moreover, the binding region of both protein and phosphopeptide is able to show more rigid and flexible while interacting with polar and non-polar partners, respectively. The example provides insight into recognition in phosphopeptides with different features.

3.3.3 Modelling the phosphorylation effect on diffusional association using coarse-grained BD simulations

Long-range electrostatic steering often plays a key role in ligand–protein association, governing the association rate constants (100–102). Therefore, it is of interest to know whether the charged phosphate group accelerates the diffusional association process of the peptide. Because of the long timescales involved in modelling the diffusion processes, simulations using classical atomistic models may be impractical, and coarse-grained models are useful tools to study molecular associations. BD simulations using a CG model were therefore used to study the effect of phosphorylation on the binding kinetics for peptides 6 and 7 from the phosphate group with and without the negative charge. Adding the phosphate group increased the net charge of the peptide by -2. Since the BRCT domain carries a -5 unit of formal charge, simulating the average association time showed no advantage in electrostatic steering attractions. Unexpectedly, although BRCT fails to form a complex with non-phosphopeptides, addition of the phosphate group demonstrates that peptides took two to three times longer to find the binding pocket and orient themselves to the final bound state. The positively charged residues in the linker region of BRCT temporarily kept the phosphopeptides locally, and sometimes the peptides stayed in a positive-charged patch longer than 15 µs. As a result, the total traffic time along the BRCT surface was increased, thus resulting in a longer BRCT–peptide association process.

It is worth noting that proteins containing the BRCT domain are significantly larger than the domain alone, which may have different charge distributions on the protein surface as well. Therefore, the example showing a slowdown in the overall BRCT–peptide association by adding the phosphate group could be an artefact of considering only the BRCT domain. Nevertheless, the results suggest that solely increasing the net charge of a molecular system may not always benefit the overall association processes, and the charge patches on a protein surface may temporarily trap the ligand. Although the phosphate group is not able to speed up phosphopeptide association, the charge interactions can efficiently guide the phosphopeptide to the correct orientation. The phosphate group stabilizes the bound BRCT–phosphopeptide complex, which is shown in both atomistic MD and CG BD simulations.

3.4. Cbl–TKB: long or short phosphopeptides?

Systematic development of short peptide or chemical inhibitors that perturb protein–protein interactions, now becoming more feasible for both cell biology and therapeutics, will benefit from modelling of peptide–protein recognition. These compounds can serve as valuable tools to dissect the complex signalling network in cellular processes and have advantages over classical biochemical techniques, such as mutagenesis, silencing RNA and gene knockouts. Typically, longer peptides provide more contacts for attraction than shorter ones. However, unusually, some truncated shorter peptides show similar or even-increasing binding affinities to their protein partner than their long peptide parent. These short peptides serve as an ideal starting point for developing inhibitors of protein–protein interactions. Because of their sizes and feasible flexibility, computational studies can be used for fundamental studies of peptide–protein recognition and as powerful tools for *in silico* peptide design.

3.4.1 Topological and 3D structures of Cbl–TKB

The casitas B-lineage lymphoma (Cbl) family proteins are evolutionarily conserved attenuators of protein tyrosine kinase signalling, whose aberrant activation is frequently associated with oncogenesis. Loss of normal Cbl functions is believed to lead to unregulated activation of regulation and cellular transformation *(103)*. Cbl proteins interact with tyrosine kinases and the binding is a phosphorylation-driven event. The N-terminal tyrosine kinase binding domain (TKB) of Cbl, Cbl–TKB, directly interacts with cognate phophotyrosyl peptide motifs of protein tyrosine kinases to regulate protein tyrosine kinase signalling *(14)*. The TKB structurally resembles an SH2-like subdomain but requires a four-helix bundle (4H) and an EF-hand subdomain to accomplish binding (Figure 3.6) *(73)*. Regulation of tyrosine kinase signalling is now widely recognized—based on inhibiting tyrosine kinase and Cbl–TKB binding with peptide inhibitors—as an oncogenesis-driving mechanism amenable to targeted therapies.

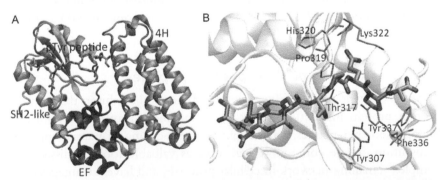

Figure 3.6 (A) TKB structure taken from PDB ID: 2CBL. (B) Zoom-in structure of interactions between phosphopeptide and Cbl–TKB domain. Phosphopeptide tail can interact with several residues from SH2-like domain. (For colour version of this figure, the reader is referred to the online version of this chapter.)

3.4.2 A twist and unusual increase in affinity from phosphopeptide truncation

Previous experiments characterized the Cbl–TKB binding site for the protein tyrosine kinase, and the domain recognizes various 12-mer phosphopeptides. The shortest one, a 10-mer phosphopeptide, has been identified from the CSF-1R protein *(74,104)*. Hydrogen binding between the phosphate group of the conserved tyrosine and nearby protein side chains plays a crucial role in binding. In addition, both termini, especially the C–terminus, form interactions with the domain. Experiments showed that Cbl–TKB can bind threefold tighter to a pentapeptide pYTPEP ($K_{eq} = 3.7$ µM) than to its parent phosphopeptide ($K_{eq} = 12.2$ µM), TLNSDGpYTPEPA identified in protein tyrosine kinase ZAP-70, where pY is phosphotyrosine *(105)*. Observing the bound state of the lead peptide inhibitor suggested that the pentapeptide adopts a poly-L-protein type II (PPII) helix; thus the successful binding may be related to the consequences of conformational constraint. Therefore, hypothesizing that a peptide with a pre-organized PPII conformation should be favourable, a new pYTP(ptE)P peptide was synthesized and tested there. In this designed peptide, glutamic acid (E) is replaced by an unnatural amino acid termed ptE and it is estimated that this pre-organization of both the backbone and side chain to the bound conformation in the pYTP(ptE)P peptide could result in nearly an order of magnitude increase in activity. NMR studies for the ptE containing peptide showed that the P(ptE)P region adopted the PPII conformation. However, biochemical assays showed that the new pYTP(ptE)P peptide lost over an order of magnitude of binding affinity with Cbl–TKB.

MD simulations were used to investigate why the designed preorganized and rigid short peptide unexpectedly failed to bind more strongly than the lead peptide. Studies of the dynamics and energy calculations show that peptide binding to Cbl–TKB involves induced fit. To achieve the final bound conformation induced by peptide binding, the protein and peptide undertake multiple dihedral rotations. In addition, the final bound state is not limited to only one conformation similar to existing crystal structures. The modelling work shows that ptE residues are exposed to solvent and may orient away from productive TKB–Cbl interaction, which is the likely cause for the observed loss of activity (Figure 3.7). As a result, overrigidifying both backbone and side-chain motions in the pYTP(ptE)P peptide reduces the local flexibility in the bound complex and introduces more unfavourable conformational changes in order to adopt the rigid peptide. Experiments in this phosphopeptide-domain system argue that the rules

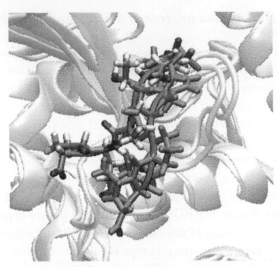

Figure 3.7 An MD snapshot of Cbl–TKB and peptide 11 (coloured) and 12 (yellow). (For interpretation of the references to colour in this figure legend, the reader is referred to the online version of this chapter.)

governing the design of potent inhibitors for interrupting protein–protein interactions are not yet fully defined. Due to the small number of examples reported in the literature, we highlight the lack of generality of these rules that scientists may rely upon to design good peptide inhibitors.

4. CONCLUDING REMARKS

Understanding phosphopeptide–modular domain recognition is a key to design peptides or peptidomimetic compounds to manipulate protein function involved in various signal transduction cascades. This review focuses on promiscuous and weak-binding protein modular domains that mediate signal transduction. These systems are different from most protein–ligand binding systems used for drug design; most drugs need to bind specifically and tightly to their target proteins. Moreover, weak binding may allow a system to be more flexible and respond faster, which has biological significance such as in DNA checkpoint signalling. From the literature, modelling tools such as MD simulations are widely applied to study the phosphopeptide–domain recognition. The atomistic detailed modelling work computed binding energies, investigated conformational changes of the molecules and modelled phosphopeptide association processes. However, because of the complex nature and highly flexible systems, the rules determining successful design of phosphopeptides are not fully identified yet. Further studies, including more thoroughly understanding of binding mechanisms, efficient conformational searches and accurate binding affinities prediction will assist structure-based phosphopeptide design.

ACKNOWLEDGEMENTS

We thank support from the National Science Foundation (MCB-0919586) and Steven Ahrendt for valuable comments on the manuscript.

REFERENCES

1. Seet, B. T.; Dikic, I.; Zhou, M.-M.; Pawson, T. *Nat. Rev. Mol. Cell Biol.* **2006**, 7, 473–483.
2. Westheimer, F. H. *Science* **1987**, 235, 1173–1178.
3. Yaffe, M. B.; Cantley, L. C. *Nature* **1999**, 402, 30–31.
4. Yaffe, M. B.; Smerdon, S. J. *Structure* **2001**, 9, R33–R38.
5. Yaffe, M. B.; Smerdon, S. J. *Annu. Rev. Biophys. Biomol. Struct.* **2004**, 33, 225–244.
6. Jorgensen, C.; Linding, R. *Brief. Funct. Genomic. Proteomic.* **2008**, 7, 17–26.
7. Chang, Chia-en. A.; McLaughlin, W. A.; Baron, R.; Wang, W.; McCammon, J. A. *Proc. Natl. Acad. Sci. U. S. A.* **2008**, 105, 7456–7461.
8. Nachman, J.; Gish, G.; Virag, C.; Pawson, T.; Pomes, R.; Pai, E. *Plos One* **2010**, 5, e11215.
9. Virshup, D. M.; Shenolikar, S. *Mol. Cell* **2009**, 33, 537–545.
10. Diella, F.; Haslam, N.; Chica, C.; Budd, A.; Michael, S.; Brown, N. P.; Trave, G.; Gibson, T. J. *Front. Biosci.* **2008**, 13, 6580–6603.
11. Mohammad, D. H.; Yaffe, M. B. *DNA Repair (Amst.)* **2009**, 8, 1009–1017.
12. Pawson, T.; Nash, P. *Science* **2003**, 300, 445–452.
13. McLaughlin, W. A.; Hou, T. J.; Wang, W. *J. Mol. Biol.* **2006**, 357, 1322–1334.
14. Lupher, M. L.; Reedquist, K. A.; Miyake, S.; Langdon, W. Y.; Band, H. *J. Biol. Chem.* **1996**, 271, 24063–24068.
15. Karplus, M.; McCammon, J. A. *Nat. Struct. Biol.* **2002**, 9, 646–652.
16. Hou, T. J.; Wang, J. M.; Li, Y. Y.; Wang, W. *J. Chem. Inf. Model.* **2011**, 51, 69–82.
17. Gan, W. X.; Roux, B. *Proteins Struct. Funct. Bioinform.* **2009**, 74, 996–1007.
18. Hou, T. J.; Zhang, W.; Case, D. A.; Wang, W. *J. Mol. Biol.* **2008**, 376, 1201–1214.
19. McClendon, C. L.; Friedland, G.; Mobley, D. L.; Amirkhani, H.; Jacobson, M. P. *J. Chem. Theor. Comput.* **2009**, 5, 2486–2502.
20. Ai, R.; Fatmi, M. Q.; Chang, C. E. A. *J. Comput. Aided Mol. Des.* **2010**, 24, 819–827.
21. Narayanan, A.; Jacobson, M. P. *Curr. Opin. Struct. Biol.* **2009**, 19, 156–163.
22. Joughin, B. A.; Tidor, B.; Yaffe, M. B. *Protein Sci.* **2005**, 14, 131–139.
23. Jorgensen, W. L.; Maxwell, D. S.; TiradoRives, J. *J. Am. Chem. Soc.* **1996**, 118, 11225–11236.
24. Okur, A.; Strockbine, B.; Hornak, V.; Simmerling, C. *J. Comput. Chem.* **2003**, 24, 21–31.
25. Hornak, V.; Abel, R.; Okur, A.; Strockbine, B.; Roitberg, A.; Simmerling, C. *Proteins Struct. Funct. Bioinform.* **2006**, 65, 712–725.
26. Vanommeslaeghe, K.; Hatcher, E.; Acharya, C.; Kundu, S.; Zhong, S.; Shim, J.; Darian, E.; Guvench, O.; Lopes, P.; Vorobyov, I.; MacKerell, A. D., Jr. *J. Comput. Chem.* **2010**, 31, 671–690.
27. Homeyer, N.; Horn, A. H. C.; Lanig, H.; Sticht, H. *J. Mol. Model.* **2006**, 12, 281–289.
28. Barducci, A.; Bonomi, M.; Parrinello, M. *Wiley Interdiscip. Rev. Comput. Mol. Sci.* **2011**, 1, 826–843.

29. Larsson, P.; Hess, B.; Lindahl, E. *Wiley Interdiscip. Rev. Comput. Mol. Sci.* **2011**, *1*, 93–108.
30. Wang, Y.; Harrison, C. B.; Schulten, K.; McCammon, J. A. *Comput. Sci. Discov.* **2011**, *4*, 015002.
31. Hamelberg, D.; Mongan, J.; McCammon, J. A. *J. Chem. Phys.* **2004**, *120*, 11919–11929.
32. Punta, M.; Coggill, P. C.; Eberhardt, R. Y.; Mistry, J.; Tate, J.; Boursnell, C.; Pang, N.; Forslund, K.; Ceric, G.; Clements, J.; Heger, A.; Holm, L.; Sonnhammer, E. L. L.; Eddy, S. R.; Bateman, A.; Finn, R. D. *Nucleic Acids Res.* **2012**, *40*, D290–D301.
33. Sammut, S. J.; Finn, R. D.; Bateman, A. *Brief Bioinform.* **2008**, *9*, 210–219.
34. Arnold, K.; Bordoli, L.; Kopp, J.; Schwede, T. *Bioinformatics* **2006**, *22*, 195–201.
35. Kiefer, F.; Arnold, K.; Kunzli, M.; Bordoli, L.; Schwede, T. *Nucleic Acids Res.* **2009**, *37*, D387–92.
36. Peitsch, M. C. *Biotechnology* **1995**, *13*, 658.
37. Eswar, N.; Webb, B.; Marti-Renom, M. A.; Madhusudhan, M. S.; Eramian, D.; Shen, M.; Pieper, U.; Sali, A. *Comparative Protein Structure Modeling with MODELLER*; John Wiley & Sons, Inc., 2006.
38. Huber, G. A.; McCammon, J. A. *Comput. Phys. Commun.* **2010**, *181*, 1896–1905.
39. Mereghetti, P.; Gabdoulline, R. R.; Wade, R. C. *Biophys. J.* **2010**, *99*, 3782–3791.
40. Frembgen-Kesner, T.; Elcock, A. H. *Biophys. J.* **2010**, *99*, L75–L77.
41. Gorecki, A.; Szypowski, M.; Dlugosz, M.; Trylska, J. *J. Comput. Chem.* **2009**, *30*, 2364–2373.
42. Kim, Y. C.; Hummer, G. *J. Mol. Biol.* **2008**, *375*, 1416–1433.
43. Carmichael, S. P.; Shell, M. S. *J. Phys. Chem. B* **2012**, *116*, 8383–8393.
44. Zheng, W. *J. Chem. Phys.* **2012**, *136*, 155103.
45. Trylska, J.; Tozzini, V.; Chang, C. A.; McCammon, J. A. *Biophys. J.* **2007**, *92*, 4179–4187.
46. Moussavi-Baygi, R.; Jamali, Y.; Karimi, R.; Mofrad, M. R. K. *PLoS Comput. Biol.* **2011**, *7*, 16.
47. Lu, L. Y.; Izvekov, S.; Das, A.; Andersen, H. C.; Voth, G. A. *J. Chem. Theor. Comput.* **2010**, *6*, 954–965.
48. Chu, J. W.; Izveko, S.; Voth, G. A. *Mol. Simul.* **2006**, *32*, 211–218.
49. Davtyan, A.; Schafer, N. P.; Zheng, W.; Clementi, C.; Wolynes, P. G.; Papoian, G. A. *J. Phys. Chem. B* **2012**, *116*, 8494–8503.
50. Clementi, C. *Curr. Opin. Struct. Biol.* **2008**, *18*, 10–15.
51. Samiotakis, A.; Homouz, D.; Cheung, M. S. *J. Chem. Phys.* **2010**, *132*, 175101.
52. Gopal, S. M.; Mukherjee, S.; Cheng, Y. M.; Feig, M. *Proteins Struct. Funct. Bioinform.* **2010**, *78*, 1266–1281.
53. Savelyev, A.; Papoian, G. A. *Biophys. J.* **2009**, *96*, 4044–4052.
54. Kamerlin, S. C. L.; Vicatos, S.; Dryga, A.; Warshel, A. In: Leone, S. R.; Cremer, P. S.; Groves, J. T.; Johnson, M. A., Eds.; Annual Review of Physical Chemistry, Vol. 62; Annual Reviews: Palo Alto, **2011**; pp 41–64.
55. Wang, Q.; Cheung, M. S. *Biophys. J.* **2012**, *102*, 2353–2361.
56. Marrink, S. J.; Risselada, H. J.; Yefimov, S.; Tieleman, D. P.; de Vries, A. H. *J. Phys. Chem. B* **2007**, *111*, 7812–7824.
57. Kang, M.; Roberts, C.; Cheng, Y.; Chang, C.-e. A. *J. Chem. Theor. Comput.* **2011**, *7*, 3438–3446.
58. Tozzini, V.; McCammon, J. A. *Chem. Phys. Lett.* **2005**, *413*, 123–128.
59. Wang, P.; Byeon, I. J. L.; Liao, H.; Beebe, K. D.; Yongkiettrakul, S.; Pei, D.; Tsai, M. D. *J. Mol. Biol.* **2000**, *302*, 927–940.
60. Liao, H.; Byeon, I. J. L.; Tsai, M. D. *J. Mol. Biol.* **1999**, *294*, 1041–1049.

61. Durocher, D.; Taylor, I. A.; Sarbassova, D.; Haire, L. F.; Westcott, S. L.; Jackson, S. P.; Smerdon, S. J.; Yaffe, M. B. *Mol. Cell* **2000**, *6*, 1169–1182.
62. Mahajan, A.; Yuan, C. H.; Lee, H.; Chen, E. S. W.; Wu, P. Y.; Tsai, M. D. *Sci. Signal.* **2009**, *2*, re12.
63. Pennell, S.; Westcott, S.; Ortiz-Lombardia, M.; Patel, D.; Li, J.; Nott, T. J.; Mohammed, D.; Buxton, R. S.; Yaffe, M. B.; Verma, C.; Smerdon, S. J. *Structure* **2010**, *18*, 1587–1595.
64. Leung, C. S.; Leung, S. S. F.; Tirado-Rives, J.; Jorgensen, W. L. *J. Med. Chem.* **2012**, *55*, 4489–4500.
65. Lee, S. J.; Schwartz, M. F.; Duong, J. K.; Stern, D. F. *Mol. Cell. Biol.* **2003**, *23*, 6300–6314.
66. Lee, H.; Yuan, C. H.; Hammet, A.; Mahajan, A.; Chen, E. S. W.; Wu, M. R.; Su, M. I.; Heierhorst, J.; Tsai, M. D. *Mol. Cell* **2008**, *30*, 767–778.
67. Hofmann, K.; Bucher, P. *Trends Biochem. Sci.* **1995**, *20*, 347–349.
68. Joshi, M.; Ebalunode, J. O.; Briggs, J. M. *Proteins Struct. Funct. Bioinform.* **2009**, *75*, 323–335.
69. Clapperton, J. A.; Manke, I. A.; Lowery, D. M.; Ho, T.; Haire, L. F.; Yaffe, M. B.; Smerdon, S. J. *Nat. Struct. Mol. Biol.* **2004**, *11*, 512–518.
70. Shiozaki, E. N.; Gu, L. C.; Yan, N.; Shi, Y. G. *Mol. Cell* **2004**, *14*, 405–412.
71. Varma, A. K.; Brown, R. S.; Birrane, G.; Ladias, J. A. A. *Biochemistry* **2005**, *44*, 10941–10946.
72. Shen, Y.; Tong, L. *Biochemistry* **2008**, *47*, 5767–5773.
73. Meng, W. Y.; Sawasdikosol, S.; Burakoff, S. J.; Eck, M. J. *Nature* **1999**, *398*, 84–90.
74. Tronrud, D. E.; Wen, J.; Gay, L.; Blankenship, R. E. *Photosynth. Res.* **2009**, *100*, 79–87.
75. Yuan, C. H.; Yongkiettrakul, S.; Byeon, I. J. L.; Zhou, S. Z.; Tsai, M. D. *J. Mol. Biol.* **2001**, *314*, 563–575.
76. Byeon, I. J. L.; Yongkiettrakul, S.; Tsai, M. D. *J. Mol. Biol.* **2001**, *314*, 577–588.
77. Lee, G. I.; Ding, Z. F.; Walker, J. C.; Van Doren, S. R. *Proc. Natl. Acad. Sci. U. S. A.* **2003**, *100*, 11261–11266.
78. Ali, A. A. E.; Jukes, R. M.; Pearl, L. H.; Oliver, A. W. *Nucleic Acids Res.* **2009**, *37*, 1701–1712.
79. Nott, T. J.; Kelly, G.; Stach, L.; Li, J. J.; Westcott, S.; Patel, D.; Hunt, D. M.; Howell, S.; Buxton, R. S.; O'Hare, H. M.; Smerdon, S. J. *Sci. Signal.* **2009**, *2*, ra12.
80. Huang, Y.-m. M.; Chang, C.-e. A. *BMC Biophys.* **2011**, *4*, 12.
81. Otte, L.; Wiedemann, U.; Schlegel, B.; Pires, J. R.; Beyermann, M.; Schmieder, P.; Krause, G.; Volkmer-Engert, R.; Schneider-Mergener, J.; Oschkinat, H. *Protein Sci.* **2003**, *12*, 491–500.
82. Lu, P. J.; Zhou, X. Z.; Liou, Y. C.; Noel, J. P.; Lu, K. P. *J. Biol. Chem.* **2002**, *277*, 2381–2384.
83. Verdecia, M. A.; Bowman, M. E.; Lu, K. P.; Hunter, T.; Noel, J. P. *Nat. Struct. Biol.* **2000**, *7*, 639–643.
84. Mittermaier, A. K.; Kay, L. E. *Trends Biochem. Sci.* **2009**, *34*, 601–611.
85. Boehr, D. D.; Dyson, H. J.; Wright, P. E. *Chem. Rev.* **2006**, *106*, 3055–3079.
86. Palmer, A. G.; Massi, F. *Chem. Rev.* **2006**, *106*, 1700–1719.
87. Morcos, F.; Chatterjee, S.; McClendon, C. L.; Brenner, P. R.; Lopez-Rendon, R.; Zintsmaster, J.; Ercsey-Ravasz, M.; Sweet, C. R.; Jacobson, M. P.; Peng, J. W.; Izaguirre, J. A. *PLoS Comput. Biol.* **2010**, *6*, e1001015.
88. Okazaki, K.-i.; Takada, S. *Proc. Natl. Acad. Sci. U. S. A.* **2008**, *105*, 11182–11187.
89. Koshland, D. E. *Proc. Natl. Acad. Sci. U. S. A.* **1958**, *44*, 98–104.
90. Fischer, E. *Ber. Dtsch. Chem. Ges.* **1894**, *27*, 2984–2993.

91. Lee, B. M.; Xu, J.; Clarkson, B. K.; Martinez-Yamout, M. A.; Dyson, H. J.; Case, D. A.; Gottesfeld, J. M.; Wright, P. E. *J. Mol. Biol.* **2006**, *357*, 275–291.
92. Kushwaha, P. S.; Mishra, P. C. *Int. J. Quantum Chem.* **2000**, *76*, 700–713.
93. Jorgensen, W. L. *Science* **1991**, *254*, 954–955.
94. Kar, G.; Keskin, O.; Gursoy, A.; Nussinov, R. *Curr. Opin. Pharmacol.* **2010**, *10*, 715–722.
95. Callebaut, I.; Mornon, J. P. *FEBS Lett.* **1997**, *400*, 25–30.
96. Huang, Y.-m. M.; Kang, M.; Chang, C.-e. A. *J. Phys. Chem.* B **2012**, *116*, 10247–10258.
97. Glover, J. N. M.; Williams, R. S.; Lee, M. S. *Trends Biochem. Sci.* **2004**, *29*, 579–585.
98. Williams, R. S.; Green, R.; Glover, J. N. M. *Nat. Struct. Biol.* **2001**, *8*, 838–842.
99. Watts, F. Z.; Brissett, N. C. *DNA Repair* **2010**, *9*, 103–108.
100. Gorfe, A. A.; Chang, C. E. A.; Ivanov, I.; McCammon, J. A. *Biophys. J.* **2008**, *94*, 1144–1154.
101. Held, M.; Noe, F. *Eur. J. Cell Biol.* **2012**, *91*, 357–364.
102. Elber, R. *Curr. Opin. Struct. Biol.* **2010**, *20*, 162–167.
103. Naramura, M.; Nadeau, S.; Mohapatra, B.; Ahmad, G.; Mukhopadhyay, C.; Sattler, M.; Raja, S. M.; Natarajan, A.; Band, V.; Band, H. *Oncotarget* **2011**, *2*, 245–250.
104. Ng, C.; Jackson, R. A.; Buschdorf, J. P.; Sun, Q. X.; Guy, G. R.; Sivaraman, J. *EMBO J.* **2008**, *27*, 804–816.
105. Kumar, E. A.; Yuan, Z.; Palermo, N. Y.; Dong, L.; Ahmad, G.; Lokesh, G. L.; Kolar, C.; Kizhake, S.; Borgstahl, G. E. O.; Band, H.; Natarajan, A. *J. Med. Chem.* **2012**, *55*, 3583–3587.

INDEX

Note: Page numbers followed by "*f*" indicate figures, and "*t*" indicate tables.

CUMULATIVE INDEX

Note: Page number followed by "*f*" indicate figures, and "*t*" indicate tables.

Printed and bound by CPi Group (UK) Ltd, Croydon, CR0 4YY
03/10/2024

01040427-0014